WITHDRAWN

NINETY ONE HILLS AREA. PANORAMIC VIEW OF THE LOWELL FORMATION AT TYPE LOCALITY

Depth of picture, taken from Hill B, about 6 miles. M—Morita formation. 9g—Lancha limestone with *Trigonia reesidei* and *Acila schencki*. Cienda limestone with *Kazanskyella Sinzowiella* is above. 9b—Black Knob dolomite. 9a—Black Knob quartzite with *Araucarioxylon*. 8c—Espinal grit overlying Baga limestone with *Paracanthohoplites*. White band between 8c and 8a (Quimbo dolomite) is Corta sandstone. 7m—Base of Arkill limestone (upper limit of *Tr. reesidei* zone). 7h—Chapparal sandstone with *Araucarioxylon*. 7d—Zone of *Pecten thompsoni*. 6g—Base of *Trigonia cragini* zone. 6f, 6d—Lower and middle *Tr. cragini* limestones. 6a—Sandstone with *Araucarioxylon* above upper *Tr. cragini* limestone. 5h—Conspicuous cross-bedded sandstone. 5g—Limestone and shale with *Dufrenoya justinae*. 3d—Shale with *Beudanticeras victoris* and *Acanthohoplites schucherti*. Limestone with *Immunitoceras immunitum* and several species of *Acanthohoplites* is above. Upper part of Lowell formation is beyond right limit of photograph.

The Geological Society of America
Memoir 38

LOWER CRETACEOUS STRATIGRAPHY IN SOUTHEASTERN ARIZONA

BY

ALEXANDER STOYANOW
University of Arizona, Tucson, Arizona

July 8, 1949

Made in United States of America

PRINTED BY WAVERLY PRESS, INC
BALTIMORE, MARYLAND

PUBLISHED BY THE GEOLOGICAL SOCIETY OF AMERICA
Address All Communications to The Geological Society of America
419 West 117 Street, New York 27, N. Y.

*The Memoir Series
of
The Geological Society of America
is Made Possible
Through the Bequest of
Richard Alexander Fullerton Penrose, Jr.*

CONTENTS

	Page
ABSTRACT	1
INTRODUCTION AND ACKNOWLEDGMENTS	2
CHAPTER 1. GENERAL PART	4
Review of previous work	4
Discovery of new faunas	5
Bisbee anticline	7
Lowell formation	8
General description	8
Section A-A in the Ninety One Hills	8
Analysis of the sequence in the type area	14
Geological map and panoramic view of the Ninety One Hills area	18
Methods in the field and in the laboratory	19
Mural limestone	20
Interpretation of paleontological stratigraphy	20
Two sections in the Bisbee area	23
Mural Hill Ridge section	23
Black Knob Ridge section	26
Eastward sections	27
Westward sections	29
Boundary between the Aptian and the Albian	31
Correlation with the Aptian-Albian sequence in the Caucasus	35
Correlation with the sequence in the Quitman Mountains, Texas	37
Correlation with the stratigraphic units of Texas	39
The Malone controversy	41
Malone facies	44
Relation between the Malone and the Lower Cretaceous Trigoniae	45
Remarks on ammonites	48
Outline of paleogeography	50
Appendix.—Note on Late Cretaceous strata in southeastern Arizona	58
CHAPTER 2. PALEONTOLOGICAL PART	61
Pelecypoda	61
Family Nuculidae	61
Family Parallelodontidae	63
Family Pernidae	64
Family Pteriidae	65
Family Ostreidae	66
Family Trigoniidae	67
Shell terminology of *Trigonia* adopted in this paper	67
Group of *Trigoniae Pseudo-Quadratae*	67
Groups of *Trigonia v-scripta* and *Trigonia vau*	79
Group of *Trigonia abrupta*	83
Group of *Trigonia excentrica*	86
Group of *Trigonia aliformis*	88
Family Pectinidae	90
Family Limidae	92
Family Astartidae	93
Family Caprinidae	94
Family Unicardiidae	95
Ammonoidea	95
Family Parahoplitidae	95

	Page
Subfamily Parahoplitinae	95
Subfamily Acanthohoplitinae	103
Subfamily Deshayesitinae	123
Family Desmoceratidae	127
Family Lyelliceratidae	128
Incertae sedis	129
REFERENCES	131
EXPLANATION OF PLATES 8–26	137
INDEX	157

ILLUSTRATIONS

PLATES

Plate
1. Ninety One Hills area. Panoramic view of the Lowell formation at the type locality..Frontispiece

Facing page
2. Orientation map of southeastern Arizona 6
3. Lower Cretaceous sequence in southeastern Arizona 18
4. Lower Cretaceous sequence in southeastern Arizona 19
5. Map of the Mural Hill Ridge 22
6. Lower Cretaceous sequence in southeastern Arizona 26
7. Lower Cretaceous sequence in southeastern Arizona 27
8. *Acila, Idonearca* 138
9. *Idonearca, Lima, Pteria, Gervillia* 139
10. *Gervillia, Exogyra* 140
11. *Ostrea, Caprina, Neithea* 141
12. *Trigonia* 142
13. *Trigonia* 143
14. *Trigonia* 144
15. *Trigonia, Astarte* 145
16. *Trigonia, Unicardium, Pecten* 146
17. *Pecten, Kazanskyella, Parahoplites* 147
18. *Sinzowiella, Beudanticeras* 148
19. *Acanthohoplites* 149
20. *Acanthohoplites, Immunitoceras* 150
21. *Paracanthohoplites, Colombiceras?, Dufrenoya* 151
22. *Dufrenoya?* 152
23. *Dufrenoya?, Dufrenoya, Sinzowiella?* 153
24. *Deshayesites?, Cleoniceras?* 154
25. *Deshayesites?, Acanthohoplites* (from Mangyshlak), *Colombiceras?* 155
26. *Douvilleiceras?, Stoliczkaia* 156
27. Geological map of the Ninety One Hills area .... at end of paper

FIGURES

Figure   Page
1. Bisbee anticline... 7
2. Carinae and furrows of a **Trigonia** shell 68

ABSTRACT

Discussions and interpretations presented in this paper are based essentially on the studies of the Lower Cretaceous sequence and marine fossil assemblages in southeastern Arizona. Described parahoplitan fauna (*Kazanskyella*, *Sinzowiella*) and most of associated lamellibranchs in the lower part of the Lowell formation (Pacheta-Saavedra members) are not represented elsewhere in North America. Here belongs the oldest known species of *Acila* (*A. schencki*) and many other pelecypods of extracontinental relations. The earliest known Cretaceous ammonite fauna of Texas (*Dufrenoya*, Travis Peak) occurs higher in the sequence (Cholla member). The next following acanthohoplitan fauna (Quajote member), exceptionally rich in species of Eurasian affinities, is again new for the United States. The strata above (Perilla member) have familiar fossils, like *Trigonia stolleyi* and *T. mearnsi* (*T. taffi* auctorum in part), and together with the Mural limestone, which with strata containing *Orbitolina texana* and rudistid reefs rests on the Lowell formation, are equivalent to the Glen Rose of Texas. The latest Lower Cretaceous ammonite zone of Arizona (equivalent of Grayson-Del Rio of Texas), with a prolific fauna of *Stoliczkaia*, is present in an isolated area south of Tucson.

Zonation of the revised and restricted parahoplitan and acanthohoplitan faunas, respectively below and above the *Dufrenoya justinae* zone, affords a good basis for approximate correlation with the Aptian of the Old World, and especially of the Caucasus. A more natural Aptian-Albian boundary, on paleontological grounds, is suggested between the *Immunitoceras nolani* and *Hypacanthohoplites jacobi* zones of the international standard Cretaceous sequence. Below the *H. jacobi* zone, shells with a strongly depressed or interrupted venter are not observed among adult forms of typical Parahoplitidae, whereas a marked and rapid development of species with more or less depressed venter is characteristic for that zone and the strata above. The true parahoplitan and acanthohoplitan faunas seem to terminate with the end of Aptian time (Clansayan), and apparently the natural lower boundary of the Albian is below the strata with numerous derivative parahoplitid and hoplitid species that have rapidly progressing ventral depression and interrupted ventral ribbing.

In southeastern Arizona the concurrent vertical distribution of index ammonites and Trigoniae throughout the Lower Cretaceous sequence permits an interpretation of the phylogenic development of Trigoniae of the pseudo-quadratae and the *v-scripta–vau* groups in relation to the ammonite zones. Ontogenic studies of Trigoniae show the trend of appreciable changes in the Pseudo-Quadratae from the Kimmeridgian (Malone, Texas), through the Neocomian (Quintuco-Agrio, Argentina), to the Aptian (Lowell, Arizona). On the other hand, certain Trigoniae of the *v-scripta–vau* groups, with all variability of characteristics in late Jurassic and early Cretaceous species, show remarkable morphological constancy from the Kimmeridgian (Malone, Texas; Catorce, Mexico) to the Aptian (Lowell, Arizona).

Paleogeographically, the Aptian (Gargasian-Clansayan) strata of southeastern Arizona are regarded as resulting from marginal incursions, with marked breaks in sedimentation and at different points of short distance from the international border, that came from a sea to the south. Periodical communication with the Aptian-Albian basin in Texas is indicated within the Arizona Cretaceous sequence by presence of the strata with Travis Peak and Glen Rose index fossils.

Twenty-one new species of ammonites and twenty-two of lamellibranchs are described in the paleontological part of the paper.

INTRODUCTION AND ACKNOWLEDGMENTS

The work which forms the basis of this paper began in 1936 upon discovery near Bisbee, Arizona, of the rich and for North America mostly new ammonite and lamellibranch faunas between the Morita formation and the Mural limestone, two standard stratigraphic units of the Lower Cretaceous sequence of Arizona. The definite relation of several collected assemblages of ammonites to the well-known parahoplitan and acanthohoplitan[1] faunas of the Caucasus and Transcaspian Region stimulated detailed research. Equally interesting problems were suggested by the great similarity of certain Cretaceous Trigoniae to the Jurassic species of Malone, Texas. Later examination of several Cretaceous sections in southeastern Arizona led to the discovery of a *Stoliczkaia* fauna in the Patagonia Mountains south of Tucson. The presence of a few Travis Peak and Glen Rose species within the middle part of the Lower Cretaceous sequence of Arizona allowed a correlation of these new faunas with the stratigraphic units of Texas. Vertical distribution of such ammonite genera as *Dufrenoya* and *Stoliczkaia* made it apparent that the fossiliferous Lower Cretaceous strata of southeastern Arizona range from pre-Travis Peak time to Grayson. However, a broader standard succession had to be inferred on the relation of the parahoplitids of Arizona to the European species, and an attempt was made to draw the Aptian-Albian boundary on what seemed to be a more natural foundation.

The stratigraphic and paleogeographic interpretations in the general part of the paper are essentially based on the discussions, conclusions, and postulations presented in the paleontological part. The close relation between the Arizona and Eurasian ammonites necessitated, within the scope of the paper, a critical revision of the family Parahoplitidae. The progress of the work was considerably retarded by the lack in Arizona, at that time, of European literature on the Mesozoic paleontology. With the kind help offered by Mr. F. O. Thompson, the well known amateur paleontologist of Des Moines, Iowa, it was possible to secure the greater part of needed publications through various antiquarian book stores in Europe and especially from the private library of the late Dr. J. F. Pompeckj. The rest of the indispensable literature I was able to obtain in bibliofilms promptly and accurately executed from the original copies available in Washington, D. C., by the Bibliofilm Service at the Library of the U. S. Department of Agriculture.

The field work was carried on with the aid of a generous grant made by the Geological Society of America from the Penrose Bequest, and the attached geological map (Pl. 27) was completed by 1938. A personal examination of certain American types, particularly those described by J. D. Dana, F. B. Meek, T. A. Conrad, R. T. Hill, F. W. Cragin, and Gayle Scott, appeared very desirable for the paleontological research, and especially with regard to the studies on the development and interrelations of Trigoniae. This was accomplished in the summer of 1940 when, with an additional grant from the Geological Society, I visited several museums. In this connection I am greatly indebted to many friends and colleagues. At Austin I had

---

[1] For the change of spelling *see* footnote 29.

the opportunity to examine a part of Cragin's collection from Malone through the kindness of Dr. F. W. Whitney of the University of Texas and Dr. H. B. Stenzel of the Bureau of Economic Geology of Texas. At Texas Christian University at Fort Worth I had a conference with Dr. Gayle Scott who kindly showed me his collections from the Quitman Mountains area. Another helpful discussion I had with Dr. W. S. Adkins at Houston. I spent about a month at Washington, D. C., studying collections preserved in the U. S. National Museum where I enjoyed unlimited help from Dr. J. B. Reeside, Jr., and Dr. L. W. Stephenson of the U. S. Geological Survey.

In the fall of the same year a trip to the Malone and Quitman mountains areas was arranged under the instructive guidance of Dr. Gayle Scott and in the company of Dr. J. B. Reeside, Jr., and Dr. R. W. Imlay. This trip afforded me an exceptional opportunity to examine personally the controversial collecting grounds of Malone. Later Dr. Scott led the party along the Rio Grande through the Cretaceous sections of the Quitman Mountains where I gained valuable information for the interpretation of certain problems connected with the Arizona stratigraphy.

At various times I have had the pleasure of showing the type section at Bisbee and my collections in the University of Arizona to Drs. Adkins, Reeside, and Scott during their visits to Arizona. I am under deep obligation to all geologists who maintained interest in the progress of this work. I am especially indebted to Dr. Reeside who many times in the course of my research supplied me with the casts of several type specimens, literature, and information, and with whom I had many stimulating discussions both personally and by correspondence.

Grateful acknowledgments are due Dr. G. M. Butler and Dr. T. G. Chapman, the past and the present deans of the College of Mines in the University of Arizona and directors of the Arizona Bureau of Mines, for the facilities in research. I wish to express my appreciation for the frequent aid rendered by Dr. B. S. Butler, Head of the Department of Geology and Mineralogy, Dr. E. D. Wilson, Geologist of the Arizona Bureau of Mines, Mr. R. E. Heineman, formerly Mineralogist of the same bureau, and the Staff of these institutions. Warm thanks are due my former students: Dr. C. A. Rasor, Dr. N. P. Peterson, and Mr. C. H. Sandberg for many days of helpful assistance in the field, Mr. C. A. Lee for constructing the topographic map of the Ninety One Hills area, and to two young but hard-working collectors, Burton Stein and Victor Stoyanow, for securing many important specimens. I wish to acknowledge the valued contributions to the work made by Mr. Tad Nichols, who prepared the panoramic picture of the Lowell formation at the type locality; by Mrs. R. A. Darrow, who examined the plant material collected from the Lowell formation; and by Mr. M. B. Lovelace, who measured the thickness of the Patagonia group. It is a special pleasure to express my gratitude to Mr. August Schlaudt, on whose property is located the type section of the Lowell formation, for his invariable hospitality during my frequent visits.

The manuscript was submitted for publication February 9, 1946.

CHAPTER 1. GENERAL PART

REVIEW OF PREVIOUS WORK

The standard sequence of the Lower Cretaceous strata of Arizona was first examined, and described as the Bisbee beds, by Dumble (1902, p. 696–715) who distinguished the following stratigraphic units, presented here in ascending order: (1) interbedded sands and clays with conglomerate at the base; (2) limestones and clays containing *Trigonia*, *Exogyra*, and other fossils; (3) interbedded limestones, clays, and sands, with oysters at base and *Caprotina*, and other fossils, at top; (4) alternating sands and clays.

Later Ransome (1904, p. 56) modified Dumble's general stratigraphic name to the Bisbee group and introduced four formations into the sequence. His formations, however, do not correspond to the four divisions of Dumble. Thus, he separated Dumble's lower division into the basal Glance conglomerate and the overlying unfossiliferous Morita formation. Very unfortunately Ransome missed Dumble's remarkably correct perception regarding the facial difference in the latter's units "two" and "three," and the individual nature of the former unit which, exceptionally rich in stratigraphically important ammonites and Trigoniae, and further referred to as the Lowell formation of the Bisbee group, is the principal subject of this paper. Ransome placed this unit with Dumble's unit "three" under a general name—the Mural limestone. The "fourth" unit of Dumble he named Cintura formation.

Ransome's characterizations of his four formations are briefly summarized (Ransome, 1904, 56–73):

GLANCE CONGLOMERATE: Largely reddish, rests unconformably on pre-Paleozoic and Paleozoic rocks; is bedded and usually reflects the lithologic characteristics of underlying material; thickness varies from 50 to 500 feet.

MORITA FORMATION: Made up of "uniformly alternating beds of dull-red shales and red or tawny sandstones, with occasional layers of grit and lenses of impure limestone." Ransome held that there was no sharp stratigraphic boundary between the Morita formation and the overlying Mural limestone of his interpretation. Of importance is his observation that there is in the sequence a bed of hard buff sandstone or quartzite which physiographically results in the formation of "a series of little bench-like spurs." This quartzite was selected by Ransome as the plane of separation between the "dominantly arenaceous Morita beds" and the "dominantly calcareous Mural limestone." The thickness of the Morita formation was given as 1800 feet.

MURAL LIMESTONE: Formed of two very dissimilar parts: The lower part consists of a "smooth under-cliff slope" made up of thin-bedded impure limestone and is 300 feet thick, the upper part is composed of "relatively thick-bedded and pure limestone." The latter culminates in a massive cliff-forming limestone with thickness varying from 50 to 200 feet. This massive limestone was pictorially described by Ransome:

"As one approaches Bisbee through the southern passes, particularly when the noonday glare has softened into the shadows of evening and the sculptured beauty of the hills is compensation for their barrenness, he is confronted by a prominent light-gray cliff crowning Mural Hill and stretching like a rampart along the face of the ridge northeast of town" (Pl. 4, fig. 1).

CINTURA FORMATION: Red, nodular shales with cross-bedded buff, tawny, and red sandstone with a few beds of impure limestone near base. It is 1800 feet thick.

The fossils collected by Ransome in the Bisbee quadrangle were discussed by Stanton (*in* Ransome, 1904, p. 70) as follows:

"All of the identified species in this collection occur in the Glen Rose beds of the lowest, or Trinity, division and only one of them (*Lunatia pedernalis*) is known to pass up into the lower members of the Fredericksburg division. The following is the list of species referable to the Glen Rose:

Orbitolina texana (Roemer) *Pecten stantoni* Hill
Glauconia branneri (Hill) *Trigonia stolleyi* Hill
Lunatia pedernalis (Roemer) *Trigonia* n. sp.
Ostrea sp. *Cyprina* sp.

"The following forms are suggestive of Fredericksburg horizons, possibly as high as the Edwards limestone:

Astrocoenia sp. "*Caprina*" sp. cf. *C. occidentalis* (Conrad)
Rhynchonella sp. *Turritella* sp. cf. *T. seriatim granulati* (Conrad)
Terebratella sp. *Actaeonella* sp. cf. *A. dolium* (Roemer)
Terebratula sp.

"The only forms in this list that have much weight are "*Caprina*" and *Actaeonella* which are not known in the Texas region below the Fredericksburg division.

"It is safe to conclude that the fossiliferous horizon represented by these collections corresponds in large part with the Glen Rose beds of Texas, and that possibly the upper portion is as high as the Edwards limestone; in other words, that they certainly belong to the Trinity division and possibly in part to the Fredericksburg division of the Comanche series."

Many writers followed Stanton's interpretation of the age of the Bisbee group. Burckhardt (1930, p. 151–152) considered the Glance-Morita sequence as Neocomian-Aptian, and the Mural limestone as another phase of Aptian rather than Albian transgression. Imlay (1939a, p. 1738, Table 2) tentatively correlated the Glance, Morita, and the lower member of the Mural limestone with the Trinity group of Texas (Upper Aptian–Lower Albian), assigning the upper member of the Mural and the Cintura formation to the Fredericksburg (Middle Albian).

A find of Cretaceous fossils in the Patagonia Mountains, south of Tucson, was reported by Schrader (1915, p. 53) who mentioned a fossiliferous thin-bedded arenaceous limestone about half a mile northwest of Mowry, from which were collected lamellibranchs and a fragment of ammonite, probably belonging to "*Acanthoceras*." Stanton reported that "these forms are sufficient to place the horizon in the Mesozoic, and all the species are apparently identical with unpublished forms known elsewhere in the Comanche series of the Cretaceous."

DISCOVERY OF NEW FAUNAS

While studying the literature connected with the problems of the Arizona Cretaceous geology I was impressed by Dumble's (1902, p. 704–706) statements on the similarity of the fauna of the Bisbee beds to "the fauna found in the section west of Sierra Blanca and in the vicinity of Malone, Texas," and also on the presence of limestone beds filled with Trigoniae in the lower part of the Cretaceous sequence southeast of Bisbee: "The Trigonias are of the type of *T. streeruvitzii*, Cragin, of the Trans-Pecos section of the Cretaceous." "*Trigonia steeruwitzi*" Cragin is a *nomen nudum* according to Adkins (1928, p. 119). Nevertheless, the latter statement of

Dumble, probably no more than a fleeting impression in the field by an observing geologist who had a general acquaintance with Cragin's assemblages of fossils from Malone, Texas, directly referred to the Malone Trigoniae and stimulated an investigation. The only described species of Trigonia heretofore reported by Ransome and Stanton was *Trigonia stolleyi* Hill, a Glen Rose species, which Ransome collected below the Mural limestone but high in the sequence.

In 1936, geologizing with my students in the hills along the international border near monument No. 91 (Pl. 3, fig. 1; Pl. 27), south of Bisbee, the area regarded by Ransome (1904, p. 67) as one of the two best collecting localities in the Bisbee quadrangle, I noticed the absence of gray limestone beds, which are so characteristic of the Mural limestone, within the 900–1000 feet of strata above the top of the Morita formation. These strata are equivalent to the "smooth under-cliff slope made of thin-bedded impure limestone," or the *lower part* of the "Mural limestone" of Ransome's definition, below the strata that are "relatively thick-bedded and culminating in the massive, cliff-forming limestone," *i.e.* the *upper part* of his Mural limestone, for which this name is here retained. The characteristics of the Mural limestone are: light-gray color, abundance of *Orbitolina texana* in the thinner beds, rudistid reefs in the massive beds; and its chemical composition: $CaCO_3$—89.77%, $MgCO_3$—2.09%, insolubles—8.14%.[2] The strata between the Morita formation and the Mural limestone, thus defined, constitute the Lowell formation of this paper.

Carefully collecting through the section, the party located beds replete with Trigoniae, some of which closely resembled the Trigoniae described by Cragin (1905) from the Malone area of Texas, undoubtedly a material similar to that noticed in the field by Dumble 34 years before. An almost immediately following find, in the near-by strata, of large ammonites of rather indefinite relationship led to the erroneous opinion that here was a sequence parallel to the "Malone and Torcer beds" of Texas (Adkins, 1932, p. 254–257, 286; Stoyanow, 1936, p. 328). Very soon, however, I located well-preserved parahoplitan faunas below the strata with the Trigoniae of Malone affinities, and an acanthohoplitan fauna above these strata. Though different from the related Caucasian species familiar to me, these ammonites definitely established the Aptian age of the Lowell formation (Stoyanow, 1938, p. 117).

Ransome did not collect ammonites within the Bisbee group. His only reference to ammonites in the Bisbee area reads: "Part of a well-preserved ammonite in the possession of Mr. W. G. McBride, of Bisbee, and said to have been picked up in Brewery Gulch, may have come from some portion of the Mural limestone exposed at the head of that gulch" (Ransome, 1904, p. 68).

The discovery of a *Stoliczkaia* fauna in the Patagonia Mountains and its relation to the Cretaceous sequence of Arizona are discussed under the heading "Westward sections."

Subsequent research in the field allowed me to establish for southeastern Arizona a stratigraphic succession based on the ammonites.

---

[2] The chemical composition may vary. The analyzed samples were taken from the quarry at Paul Spur (see Pl. 2) between Bisbee and Douglas: Lausen, C. and Wilson, E. D., "Cement investigations," unpublished manuscript, 1925.

BISBEE ANTICLINE

The crescentic and pitching Bisbee anticline was not discussed by Ransome in his outline of the geological structure of the Bisbee area except to suggest that a northeastward tilting of the area may have resulted from "a broad anticlinal structure" of which the Mule Mountains, composed of Cretaceous strata, is the northeastern

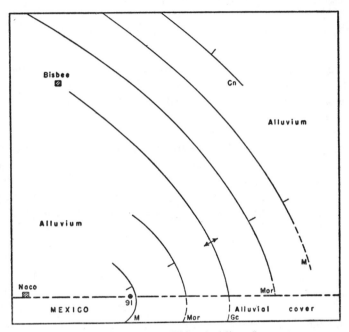

FIGURE. 1.—*Bisbee Anticline*

Gc—Glance conglomerate. Mor—Morita formation. M—Lowell formation overlain by Mural limestone. Cn—Cintura formation.

flank, the southwestern flank having been destroyed by downfaulting or erosion (Ransome, 1904, p. 107–108). The possible existence of a buried southwestern flank of the Cretaceous anticline, which probably was fairly symmetrical, farther southeastward in the area, was also briefly mentioned in connection with a local overthrust (Ransome, 1904, p. 102). Later this fold was discussed by Tenney (1932, p. 49–51, Fig. 3).

The arching of the basal Glance conglomerate (Pl. 7, fig. 2) and the successive dipping away from the axis, in the symmetrically opposite directions, of the Morita formation, the Lowell formation, and the Mural limestone, are observable in the field north of the international border (Fig. 1) and are clearly indicated in Ransome's geological map of the Bisbee quadrangle (Ransome, 1904, Pl. 1, section E-E). The anticline pitches southeastward, and it is apparent that in the vicinity of the international border its axis curves so as to place the Ninety One Hills (Pl. 3, fig. 1; Pl. 4, fig. 2), on which the international boundary monument No. 91 is located, at the focal point of the crescent.

The type locality of the Lowell formation, which is described in the following pages, is in the southwestern flank of the anticline, southeast of Bisbee and east of Naco (Pl. 2, 1). The other two discussed sections across the Mule Mountains: the Mural Hill Ridge section (Pl. 2, 3) and the Black Knob Ridge section (Pl. 2, 2) are in the northeastern flank of the anticline. South of the international border the pitching anticline seems to be deeply buried in the alluvium.

LOWELL FORMATION
GENERAL DESCRIPTION

In the extreme southeastern part of Arizona the Lower Cretaceous sequence is best exemplified in the Bisbee area, and its lower fossiliferous division—the Lowell formation of the Bisbee group—is best represented in the Ninety One Hills, south of Bisbee Junction and north of the international border (Pl. 27). To the north, in the Bisbee mining district and in the Mule Mountains, several units of the Lowell formation grade laterally into silicified limestones and quartzites, and are partially altered by the igneous agencies and mineralization. Immediately to the south, in Mexico, very little of the formation is observable because of a thick alluvial cover, although undoubtedly it is exposed again below the Mural limestone farther south of the international border in many mountainous ridges of a general southeasterly trend. It has not been found west of Bisbee. To the east, as far as the New Mexico state line, only the upper members of the formation are identifiable; its major part is represented by the lithologically different rocks with poorly preserved and probably not directly related faunas.

In the unaltered beds of the Lowell formation of the type area the fauna is very abundant and well preserved, but only the stratigraphically important ammonites and lamellibranchs have been studied for the purpose of this research. The succession of the strata and their index fossils, as it appears in the standard section of the Lowell formation in the Ninety One Hills along line A-A of the geological map (Pl. 27), is presented below. The index fossils are described and illustrated in the paleontological part of this paper. This section has been selected because within it there is no structural interference with the normal succession of the strata between the Morita formation and the Mural limestone.

The members of the Lowell formation are so chosen as to represent definite units of faunal assemblages which naturally coincide, for the major part of members, with the regimen of the sea and shore line. Special care was taken to draw the boundaries between the members at the lithologically and topographically most conspicuous key beds (Pls. 1, 27).

SECTION A-A IN THE NINETY ONE HILLS

Top   Thickness in feet

MURAL LIMESTONE. Light-gray limestone, in the lower part bedded, with *Orbitolina texana* (Roemer) and *Lima muralensis*, sp. nov. (Pl. 9, fig. 2); in the upper part massive, with rudistid reefs. Surface eroded.

LOWELL FORMATION
    Pedregosa member. Division 1a. Thin layers of light-gray argillaceous limestone, friable green sandstone, and crumbly light-gray to purplish shale.................. 30

LOWELL FORMATION

Top	Thickness in feet

    b. Friable gray, white, and yellow shale and greenish sandstone.. | 105
    c. At base, layers of light-buff calcareous sandstone with molds of lamellibranchs. Rest of unit made up of buff, yellow, and greenish friable sandstone in layers a few inches thick..................................... | 25
    d. Gray, in places buff, limestone replete with small turritelloid shells. Specimens of *Nerinea* sp.. | 1
    e. White platy sandstone with small turritelloid shells. | 20
    f. Conspicuous buff sandstone weathering smooth.... | 1
                                                         Total........ | 182

Perilla member. Division 2a. Brown crystalline limestone, very fossiliferous (lamellibranchs)... | 5
    b. Friable light-yellow shale and disintegrated argillaceous limestone. Mostly covered with detritus. *Trigonia mearnsi*, sp. nov. (Pl. 12, figs. 5–6; Pl. 13, figs. 1–4), mainly in the lower part of the unit but abundantly represented for a considerable distance upward. *T. stolleyi* Hill (Pl. 15, figs. 9–11), *Astarte adkinsi*, sp. nov. (Pl. 15, figs. 12–14), *Neithea vicinalis*, sp. nov. (Pl. 11, figs. 6–7), *Pteria peregrina*, sp. nov. (Pl. 9, fig. 5)... | 85
    c. Very fossiliferous (lamellibranchs) brown crystalline limestone.. | 1
    d. Friable light-yellow shale with occasional layers of brown and yellow limestone with oyster banks.................................. | 160
                                                         Total........ | 251

Quajote member. Division 3a. Light-buff streaked sandstone..................... | 1
    b. Hard brownish limestone, greenish-gray on the fresh fracture, replete with lamellibranchs and ammonites. *Immunitoceras immunitum*, gen. and sp. nov. (Pl. 20, figs. 8–15), *Acanthohoplites berkeyi*, sp. nov. (Pl. 19, figs. 14–16), *A. erraticus*, sp. nov. (Pl. 19, figs. 21–23), *A. hesper*, sp. nov. (Pl. 20, figs. 1–6), *A. impetrabilis*, sp. nov. (Pl. 19, figs. 17–20), *A. teres*, sp. nov. (Pl. 20, fig. 7), *Unicardium* sp. (Pl. 16, fig. 4).. | 2
    c. Weak drab fossiliferous limestone................... | 1
    d. Whitish and yellow-green shale, and greenish sandstone with layers of dark-green limestone which is replete with small lamellibranchs. In the concretions and local irregular accumulations of greenish limestone occur ammonites: *Acanthohoplites schucherti*, sp. nov. (Pl. 19, figs. 1–13), *Beudanticeras victoris*, sp. nov. (Pl. 18, figs. 18–21), and *Colombiceras? brumale*, sp. nov. (Pl. 21, figs. 8–10). In light-colored shales and in layers of dark-green limestone are numerous specimens of *Trigonia weaveri*, sp. nov. (Pl. 15, figs. 4–8), *Grammatodon* sp., and pseudo-quadrate Trigoniae related to the species of the Cholla member........................... | 11
    e. Fossiliferous friable yellow sandstone with a bed of hard buff sandstone at the top... | 11
    f. Fossiliferous friable yellow sandstone alternating with layers of greenish-yellow sandstone.. | 6
                                                        Total........ | 32

Division 4a. Brown and pinkish sandstone weathering into gray angular pieces, with layers and lenses of fossiliferous limestone.................... | 10
    b. Friable gray and yellow shale capped by thin-bedded and platy whitish sandstone.. | 2
    c. Thin-bedded yellow and gray shale grading upward into harder yellow and brown sandstone.. | 3

Top | Thickness in feet

   d. Hard thin layers of yellow crystalline limestone replete with lamellibranch shells. Black spots on the bedding planes from numerous fragments of oysters... 3

   e. Friable light-yellow sandstone, harder, platy, and calcareous toward the top... 3

                Total........ 21

Cholla member. Division 5a. Massive streaked and cross-bedded sandstone........ 1

   b. Platy, light-colored, white, yellow, and brownish sandstone with layers of gray limestone which weathers into medium-sized balls........ 13

   c. Streaked and cross-bedded pinkish sandstone........ 2

   d. Friable light-yellow sandstone, harder, platy, and calcareous toward the top.. 6

   e. Gray limestone with ammonites and echinoids (Spatangidae). Upper limit of *Trigonia kitchini*, sp. nov. (Pl. 14, figs. 4–10). *Lima cholla*, sp. nov. (Pl. 9, figs. 3–4)... 3

   f. Crumbly yellow shale with fragments of ammonites from D.5g........ 9

   g. Light-buff limestone weathering round. Oysters, large ammonites. *Dufrenoya justinae* (Hill) (Pl. 21, figs. 11–17), *D.? compitalis*, sp. nov. (Pl. 22, figs. 1–8; Pl. 23, figs. 1–3)................................ 3

   h. Conspicuous pinkish, gray on the fresh fracture, cross-bedded and streaked sandstone.. 2

   i. Platy thin-bedded shale........................... 8

   k. Buff streaked sandstone.......................... 2

   l. White shaly sandstone with numerous fragments of silicified wood... 19

                Total........ 68

   Division 6a. Hard light-buff to yellow limestone weathering into irregularly rounded boulders (Pl. 7, fig. 1) with "*Astrocoenia*" and "*Serpula*" in abundance. Toward the international border this limestone becomes somewhat arenaceous. Numerous accumulations of specimens of *Trigonia kitchini* (lower limit), *T. cragini*, sp. nov. (Pl. 13, figs. 6–10; Pl. 14, fig. 3. Upper limit of this species), *T. guildi*, sp. nov. (Pl. 12, figs. 1–2), and *T. resoluta*, sp. nov. (Pl. 12, fig. 3). Poorly preserved ammonites and gastropods. The hard limestone is underlain by thin crumbly shaly limestone with oysters... 3

   b. Hard pink cross-bedded sandstone.................. 3

   c. White friable shaly sandstone and shale with fragments of silicified wood.. 33

   d. Gray arenaceous limestone which weathers into round boulders similar to limestone 6a. *Trigonia cragini* and pseudo-quadrate Trigoniae as in 6a. *Gervillia cholla*, sp. nov. (Pl. 10, fig. 1), *G. rasori*, sp. nov. (Pl. 10, figs. 2–3) 2

   e. Soft shaly and platy sandstone with remains of silicified trees... 22

   f. Gray argillaceous limestone similar to 6a and 6d, and with the same species of pseudo-quadrate Trigoniae. Lower limit of *T. cragini*. A specimen of *Acila (Truncacila)*, belonging to a species different from *A. schencki* of the Lancha limestone at the base of the formation, was collected from this stratum...... 2

   g. Pink streaked sandstone with distinct bedding........ 5

   h. Detritus between 6g and 6i....................... 9

   i. Soft bedded sandstone........................... 32

LOWELL FORMATION

	Thickness in feet
Top	

k. Light-gray mottled and conglomeratic limestone with fragments of oysters and inclusions of yellow dolomite. A very conspicuous and widespread bed.. 1

Total........ 112

Saavedra member. Division 7a. Slope-forming bedded light-colored calcareous sandstone.. 27
　　　　　　　　b. Ridge-forming, very platy and thin-bedded, light-brown sandstone. Darker speckled sandstone at the top........................ 11
　　　　　　　　c. White, green, and purple soft shale............... 25
　　　　　　　　d. Argillaceous limestone, mottled and grayish on the fresh fracture, weathers angular and light-gray, replete with shells of *Pecten* (*Chlamys*) *thompsoni*, sp. nov. (Pl. 16, figs. 5-8; Pl. 17, fig. 1) and *Ostrea edwilsoni*, sp. nov. (Pl. 11, figs. 1-3). Poorly preserved gastropods..................................... 5
　　　　　　　　e. Green shale................................... 4
　　　　　　　　f. Dense gray limestone weathering brownish and round. Varietal forms of *P. thompsoni*. Poorly preserved ammonites. Numerous gastropods and worms.. 2
　　　　　　　　g. Light-colored shales and sandstones, reddish at the base... 25
　　　　　　　　h. Chapparal sandstone. White saccharoidal sandstone (pinkish or rusty in other sections) weathering into round-edged slabs. Large trunks of *Araucarioxylon*.. 24
　　　　　　　　i. Very fossiliferous yellow limestone, contains numerous specimens of unusually large lamellibranchs and gastropods. Pseudo-quadrate and scabroid Trigoniae. *Gervillia heinemani*, sp. nov. (Pl. 9, figs. 6-7)............ 2
　　　　　　　　k. Soft white, in places buff, sandstone............... 14
　　　　　　　　l. Arkill limestone. Grayish-brown and buff arenaceous limestone replete with shells of *Trigonia reesidei*, sp. nov. (Pl. 14, figs. 11-14; Pl. 15, figs. 1-3). Upper limit of this species. Poorly preserved ammonites......... 2
　　　　　　　　m. Buff sandstone with whitish grains............... 12
　　　　　　　　n. Barata limestone. Streaked calcareous sandstone near the top and fossiliferous limestone below. Lower limit of pseudo-quadrate Trigoniae. *Trigonia saavedra*, sp. nov. (Pl. 13, fig. 5). Numerous gastropods and lamellibranchs... 2
　　　　　　　　o. Drab, gray arenaceous limestone................. 6
　　　　　　　　p. Light-brown sandstone......................... 7
　　　　　　　　s. White sandstone............................... 40
　　　　　　　　t. Detritus from the above unit, approximately...... 10

Total........ 218

Joserita member. Division 8a. Quimbo dolomite. Ochre-colored dolomite, in places grades laterally into light-yellow calcareous shale................................. 3
　　　　　　　　b. Corta sandstone. Hard cross-bedded, partly quartzitic and partly calcareous, reddish-brown and in places whitish sandstone.......... 35
　　　　　　　　c. Espinal grit (Pl. 6, fig. 1). Coarse, partly conglomeratic, grit with imperfectly rounded and subangular pebbles of limestone, quartzite, chert, arkose, and sandstone. Large inclusions of the underlying yellow dolomite 8d. Toward the international border the grit grades into sandstone, arenaceous shale, and limestone, the latter buff on the surface, greenish on the fresh fracture. All these rocks contain pebbles in variable amounts. Large ammonites, Trigoniae, Limae, pectens,

	Thickness in feet
Top	
oysters. *Deshayesites? butleri*, sp. nov. (Pl. 25, figs. 1–3), *Lima espinal*, sp. nov. (Pl. 9, fig. 8), *Trigonia reesidei*. Average thickness.................................	4
d. Yellow dolomite with numerous ammonites and lamellibranchs. Lowest observed limit of *Dufrenoya*. *Dufrenoya joserita*, sp. nov. (Pl. 23, figs. 4–6)...	3
e. Baga shale and limestone. Dark-gray and greenish impure limestone in places lens-shaped and grading laterally into white shale. Numerous lamellibranchs. *Paracanthohoplites meridionalis*, gen. and sp. nov. (Pl. 21, figs. 1–7)..	7
f. Mostly soft shale and sandstone...................	35
g. Gray shaly limestone with oysters.................	2
h. Soft sandstone and fine-grained arenaceous shale in the upper part, maroon shale in the lower part................................	34
Total........	123

Pacheta member. Division 9a. Black Knob quartzite. Greenish quartzite in places overlain with buff and white speckled sandstone. Large silicified trunks of *Araucarioxylon* and sporadic shells of *Analina* sp. Between the quartzite and the underlying dolomite 9b is a whitish-gray limestone into which the dolomite probably grades. In places the limestone encroaches upward at the expense of the quartzite............ 40

    b. Black Knob dolomite. Yellow dolomite, pinkish when fresh, in places silicified. Reptilian remains.. 30

    c. Cienda limestone. Minutely clastic argillaceous limestone with glauconite in the matrix and with numerous larger inclusions of broken shell and chert. Weathers into round-edged slabs. Very rich in silicified ammonites (no other fossils observed). *Kazanskyella arizonica*, gen. and sp. nov. (Pl. 17, figs. 2–8), *Sinzowiella spathi*, gen. and sp. nov. (Pl. 18, figs. 1–17)..................... 4

    d. Tusonimo limestone. Gray weak arenaceous limestone, in places pinkish and "earthy," with fragments of lamellibranchs and poorly preserved specimens of ammonites. *Sinzowiella?* sp. (Pl. 23, figs. 7–13)............ 8

    e. Streaked and cross-bedded sandstone with fragments of silicified trees... 2

    f. Yellow dolomite................................. 2

    g. Lancha limestone. Grayish, bluish-gray, and pinkish limestone in places arenaceous or partly silicified. In upper layers contains ammonites related to *Kazanskyella*. Numerous silicified shells of *Trigonia reesidei*, *Acila (Truncacila) schencki*, sp. nov. (Pl. 8, figs. 1–6), *Idonearca stephensoni*, sp. nov. (Pl. 8, figs. 9–12; Pl. 9, fig. 1), *Exogyra lancha*, sp. nov. (Pl. 10, figs. 4–6), and *Anomia* sp. Abundant small long-spired gastropods with variable ornamentation........................ 11

    Total........ 97

Total thickness of the Lowell formation, as measured............................ 1104

MORITA FORMATION. Maroon mudstone, shale, and speckled quartzite.

    The contact between the Morita formation and the Lancha limestone is very conspicuous in the field because of the marked difference in rock composition and color, but here and there in the upper layers of the Morita are observed sudden discolorations of the maroon mudstone and shale, and their replacement by white argillaceous limestone. In the Ninety One Hills area I have seen only the angular inclusions of the Morita material in the limy rock, but not vice versa. To the north, however, in

LOWELL FORMATION

Table 1.—*Distribution of index species in the Lowell formation*

Described species from the Lowell formation	Member	Perilla	Quajote		Cholla					Saavedra				Joserita			Pacheta		
		2b	3b	3d	5e	5g	6a	6d	6f	7d	7i	7l	7n	8c	8d	8e	9c	9d	9g
Acila (Truncacila) schencki																				X
Acanthohoplites berkeyi			X																	
Acanthohoplites erraticus			X																	
Acanthohoplites hesper			X																	
Acanthohoplites impetrabilis			X																	
Acanthohoplites schucherti				X																
Acanthohoplites teres			X																	
Astarte adkinsi		X																		
Beudanticeras victoris			X																	
Colombiceras? brumale				X																
Deshayesites? butleri														X						
Dufrenoya joserita															X					
Dufrenoya justinae					X															
Dufrenoya? compitalis					X															
Exogyra lancha																				X
Gervillia cholla									X											
Gervillia heinemani												X								
Gervillia rasori									X											
Idonearca stephensoni																				X
Immunitoceras immunitum			X																	
Kazanskyella arizonica																	X			
Lima cholla					X															
Lima espinal															X					
Neithea vicinalis		X																		
Ostrea edwilsoni											X									
Paracanthohoplites meridionalis																X				
Pecten (Chlamys) thompsoni											X									
Pteria peregrina		X																		
Sinzowiella spathi																X				
Sinzowiella? sp.																		X		
Trigonia sp. ex aliformis gr.		X																		
Trigonia cragini								X	X	X										
Trigonia guildi									X											
Trigonia kitchini						X			X											
Trigonia mearnsi		X																		
Trigonia reesidei														X		X				X
Trigonia resoluta									X											
Trigonia saavedra															X					
Trigonia stolleyi		X																		
Trigonia weaveri				X																
Unicardium sp.				X																

the Mule Mountains, the contact between the Morita formation and the basal limestones of the Lowell formation is not so sharp: there is a zone of alternating maroon shales and limestone beds there, and the angular inclusions of limestone occur in the shale.

The conditions of deposition of the Lowell formation were oscillatory. Within the sequence six stratigraphic units with silicified tree remains (Divisions 9e, 9a, 7h, 6e, 6c, and 5l) are separated one from another by fossiliferous marine strata of divers thicknesses. The tree remains range from scattered fragments of various sizes to large tree trunks (Black Knob quartzite; Chapparal sandstone). The thin sections show that all collected examples belong to *Araucarioxylon*. The oscillations were brief and minor, in an area that was in close proximity to the open sea. This is inferred from the fact that in the marine strata, which alternate with the plant-bearing beds, occur not only lamellibranchs but also ammonites, and in certain strata ammonites only. Nevertheless, it is safe to conclude that such large tree trunks as are abundant in the Black Knob quartzite and the Chapparal sandstone could not have been transported very far from the shore line and indicate the presence of a near-by forested land. In the Black Knob Ridge section several such trees are normal to the bedding planes.

Ammonites of various degree of preservation occur sporadically at many horizons of the Lowell formation, but they are especially numerous and better preserved in the limestones of the Pacheta (D.9c), Joserita (D.8e), Cholla (D.5g), and Quajote (D.3d and 3b) members. The comparative brevity of interruptions in the marine deposition within the sequence is interpreted stratigraphically from the succession of parahoplitan-dufrenoyan-acanthohoplitan faunas, which appears to be, in the main, as normal as in regions with an uninterrupted marine sedimentation. The successive ammonite faunas were evidently able to arrive in the local depositional area shortly after the intervals.

ANALYSIS OF THE SEQUENCE IN THE TYPE AREA

Pacheta member.—The Lancha limestone (D.9g) is the basal stratum of the Lowell formation. Of its index fossils, *Acila (Truncacila) schencki* probably is the oldest known representative of the genus. The age of this species is determined by the parahoplitan ammonites which occur in the upper part of the Lancha limestone and above it. The oldest species of *Acila* described to date—*Acila (Truncacila) bivirgata* (J. de C. Sowerby)—is found in the Folkestone Gault of England (Woods, 1903, p. 20) and is said to occur also in the "Aptian-Albian" strata of eastern Venezuela (Schenck, 1936, p. 48). *Acila schencki* is very abundant in the Lancha limestone, as are *Idonearca stephensoni*, *Exogyra lancha*, and *Trigonia reesidei*. *Trigonia reesidei* may be related to an undescribed form of unknown exact stratigraphic position from the Cretaceous of Texas.

The Lancha limestone is overlain by yellow dolomite (D.9f) and cross-bedded sandstone with silicified wood (D.9e). The deepening of the area and a rapid spread of marine waters over the shallow-water deposits are indicated by the reappearance of ammonites in the Tusonimo limestone (D.9d), and especially in the stratigraphically higher Cienda limestone (D.9c) which contains inclusions of the underlying yellow dolomite and is characterized by *Kazanskyella arizonica* and *Sinzowiella spathi*, two species that are closely related to the parahoplitan fauna of the Aptian strata in the Caucasian and Transcaspian regions of Russia. Another shallowing of the area followed and culminated in the formation of the Black Knob quartzite (D.9a) with large trees.

Joserita member.—The basal strata of this member (Divisions 8h, 8g, and 8f) again show a gradual reversion to the sea conditions. The first ammonite to appear in the Joserita member is *Paracanthohoplites meridionalis* of Caucasian relation which occurs in the lenses of the Baga limestone (D.8e) in association with numerous lamellibranchs. In the overlying yellow dolomite (D.8d) occurs *Dufrenoya joserita*, the oldest representative of *Dufrenoya* observed in the sequence, and numerous fragmentary specimens of other ammonite species of yet unestablished relations. The next unit—the Espinal grit (D.8c) is of special interest. Notwithstanding the proximity of a shifting shore line this partly conglomeratic stratum, with heterogeneous pebbles and inclusions of the underlying yellow dolomite, must have been formed under comparatively quiet conditions of deposition to account for a satisfactory preservation of such large ammonite shells as that of *Deshayesites? butleri* and of unbroken specimens of lamellibranchs like the collected examples of *Trigonia reesidei* and *Lima espinal*. Among the undescribed lamellibranchs of this unit the presence of Trigoniae of the *excentrica* group should be noted. No evidence of fossils has been observed in the Corta sandstone (D.8b) and the Quimbo dolomite (D.8a) which overlie the Espinal grit and compose the upper part of the Joserita member.

Saavedra member.—Upward from the base of this member fine-grained sandstones grade through calcareous sandstones and arenaceous limestones, into almost pure limestone beds. With the deposition of the Barata limestone (D.7n) a new fauna, not observed in the two lower members of the formation, and in a general way resembling the Lower Cretaceous faunas of Texas, was introduced in the area. The faunal assemblage contains numerous specimens of *Astarte, Cucullaea, Lucina, Pecten*, and large forms of *Glauconia* or *Cassiope*. The pseudo-quadrate Trigoniae, represented by *Trigonia saavedra*, and the scabroid Trigoniae make their first appearance in the sequence. Among the pectens, I found an incomplete specimen which may be related to *Pecten stantoni* Hill (Hill, 1893, p. 24, Pl. 2, figs. 3-3a). The ammonites thus far collected from the Barata limestone are not identifiable.

A buff sandstone separates the Barata limestone from the slabby arenaceous Arkill limestone (D.7l) in which *Trigonia reesidei* recurs for the last time and in this way overlaps the stratum with the first pseudo-quadrate Trigoniae of the sequence. Ammonites and Trigoniae of the *excentrica* group also are present in the Arkill limestone, but the collected material is neither abundant nor satisfactorily preserved for a study.

Fourteen feet above the Arkill limestone, in the exceptionally fossiliferous yellow limestone D.7i, are very large specimens of lamellibranchs and gastropods which remind one of certain assemblages with large forms such as those described from the Cretaceous strata of Patagonia (Belgrano beds) and the Neuquen area of Argentina (Stanton, 1901; Weaver, 1931). Especially abundant are large specimens of *Astarte, Cucullaea, Gervillia, Lucina, Perna, Unicardium, Buccinopsis*, and *Rostellaria?*

Since the large fossil trees were found in the overlying white and soft Chapparal sandstone (D.7h), the excavation of the trunks has become popular in the community, and several fairly large specimens of *Araucarioxylon* may be seen now at the auto courts and service stations of Bisbee.

About 25 feet of shales and sandstones deposited in quiet shallow waters separate the Chapparal sandstone from the limestone D.7f from which numerous but not well-preserved ammonites, gastropods, and lamellibranchs were collected. A similar but better-preserved fauna is present in the very fossiliferous and stratigraphically higher limestone D.7d with the banks of *Pecten (Chlamys) thompsoni* and *Ostrea edwilsoni* which are related to Europena species. No fossils have been found in the upper 50 feet of the Saavedra member.

Cholla member.—The lower part of the Cholla member (D.6) begins with thin but widespread conglomeratic limestone (D.6k) which is overlain by soft sandstones. The strata above these sandstones mark three rapid oscillations in the area. These are three similar and very conspicuous limestone beds (D.6f, D.6d, and D.6a), which may be called "Trigonia beds" for the abundance of Trigoniae of the pseudo-quadratae and *v-scripta–vau* groups of the Malone facies. Each of the three limestone beds is immediately succeeded by soft sandstones and shales with the remains of *Araucarioxylon* (D.6e, D.6c, and D.5l). Such species as *Trigonia guildi, T. resoluta, T. cragini, Gervillia cholla,* and *G. rasori* occur in all three limestones. A fragmentary specimen of *Acila (Trunacila)* has been collected from the lower limestone. *Trigonia kitchini*—a species with a more advanced ornamentation than that of *T. cragini*—has not been found below the upper limestone from which also a few examples of *Salenia* sp. and *Tetragramma?* sp. have been collected. Representatives of the following genera are common to all three limestone beds: *Arctica, Astarte, Clementia, Cucullaea, Cyprimeria, Gervillia* (small, ornamented forms), *Lucina, Modiola, Ostrea, Pecten* (coarse-ribbed forms), *Homomya?, Pleuromya?, Plicatula,* "*Astrocoenia,*" "*Serpula,*" and "*Vermetes.*" Ammonites occur sporadically in the Trigonia limestones, but no workable specimens have been found.

The upper division of the Cholla member (D.5) also begins with sandstones. In its middle and upper parts the sandstone and limestone beds alternate, but the latter are thinner and less conspicuous in the field than the Trigonia limestones in the lower division of the member. Stratigraphically important are the limestone D.5g and the overlying shale D.5f with *Dufrenoya justinae* and *D.? compitalis*. Other fossils in these strata are oysters and spatangoid echinoids. The gray limestone D.5e marks the upper limit of *Trigonia kitchini*. The Cholla member ends in the sandstone beds, some of which are cross-bedded.

Quajote member.—A further deepening of the depositional area is indicated already in the lower division of the Quajote member. The number of fossiliferous beds (lamellibranchs) rapidly increases in the basal strata of its upper division (D.3). With the development of lens-shaped accumulations of greenish limestone in the shale D.3d is introduced a new ammonite and lamellibranch fauna of European (*Acanthohoplites schucherti, Beudanticeras victoris*) and South American (*Colombiceras ? brumale, Trigonia weaveri*) aspects. Specimens of lamellibranchs, especially those belonging to several species of *Astarte, Grammatodon, Meretrix,* and *Unicardium,* are abundant. The pseudo-quadrate Trigoniae collected from this unit are closely related to the species of the Cholla member.

Only a one-foot limestone bed separates the unit D.3d from the unit D.3b. Although the latter is a limestone averaging not over 2 feet, the number of ammonites it contains is astonishing. I have collected from this stratum several hundred specimens at a single point of the field, and apparently an unlimited number of examples may be obtained by breaking the bed along the strike. However, the collecting is a matter of hammer, chisel, art, and luck. Under Arizona climatic conditions the weathered specimens do not come out of this hard brownish-green limestone, and much of it is buried in the alluvium and detritus, which probably accounts for the fact that previous workers missed a bed so rich in ammonites. Six representative species of this faunule are described and illustrated in the paleontological part of this paper.

All the ammonites collected from the unit D.3b belong in the subfamily Acanthohoplitinae (subf. nov.), and some of them are more or less related to the species and varieties described by Sinzow (1908) from the Mángyshlak[3] Peninsula and adjacent areas of the Transcaspian Region, Russia.

Perilla member.—The strata of the Perilla member are very fossiliferous, with many oyster banks and accumulations of other lamellibranchs, principally of the genera *Astarte, Cucullaea, Cyprimeria, Gervillia, Isocardia, Panope, Pecten* (large, coarsely ribbed forms), *Protocardia*, and *Trigonia*. The stratigraphically important unit D.2b is made up of shale and argillaceous limestone, in which *Trigonia mearnsi* and *T. stolleyi* are abundant. In the Quitman Mountains, Texas, the former species occurs with *Trinitoceras* below the *Orbitolina texana* zone. The latter species is common in the Glen Rose. Other described lamellibranchs of the Perilla member are *Astarte adkinsi, Pteria peregrina*, and *Neithea vicinalis*. In the upper part of the member, above the *T. mearnsi* zone, occur specimens of ammonites, large and small, but as yet no forms serviceable for interpretation have been found.

Pedregosa member.—In the Ninety One Hills area the upper member of the Lowell formation consists of about 200 feet of light-colored yellow, buff, brownish, and green sandstone and impure limestone beds some of which are replete with small highspired gastropods. In the type area the Pedregosa member is very distinctly separated from the basal part of the Mural limestone which begins with gray limestone layers characterized by *Orbitolina texana* and *Lima muralensis*, sp. nov. (Pl. 9, fig. 2). This sudden change from the varicolored, predominantly brownish strata of the Pedregosa to the monotonously whitish-gray beds of the Mural is quite a striking contrast in the field. Nowhere within the Lowell formation's 1100 feet of strata in the Ninety One Hills area have rocks been observed even remotely resembling the Mural limestone.

The introduction and development of the Mural limestone mark the end of the frequently shifting shore line and brief oscillations, and inaugurate a general deepening in a larger region with more or less uniform depositional conditions.

However, even within the Bisbee area, the contact between the slope-forming multicolored strata of the Lowell formation and the cliff-forming, more resistant, gray

---

[3] This is correct English spelling for Germanized "Mangyschlak." *See* "Times Atlas and Gaseteer of the World," 1922.

Mural limestone is not everywhere as definite and simple as near the international border. In the Mule Mountains, a zone below the massive limestone of the Mural contains ferruginous yellow and brown argillaceous and arenaceous beds, so characteristic of the Pedregosa member, alternated with the gray limestone layers of the

TABLE 2.—*Comparative vertical distribution of the ammonites and Trigoniae in the Lowell formation*

Member		Vertical range of Trigoniae and ammonites in the Lowell formation
Pedregosa		
Perilla	2b	*Trigonia mearnsi, Trigonia stolleyi*
Quajote	3b	*Acanthohoplites* zone. *Immunitoceras immunitum*
	3d	*Trigonia weaveri, Beudanticeras victoris, Colombiceras? brumale, Acanthohoplites schucherti*
Cholla	5e	
	5g	*Trigonia kitchini* |*Dufrenoya justinae*
	6a	*Trigonia guildi*
		Trigonia resoluta
	6d	*Trigonia cragini*
	6f	
Saavedra	7d	
	7i	
	7l	
	7n	*Trigonia saavedra*
Joserita	8c	*Deshayesites? butleri*
	8d	|*Dufrenoya joserita*
	8e	*Paracanthohoplites meridionalis*
Pacheta	9c	*Kazanskyella arizonica, Sinzowiella spathi*
	9d	*Sinzowiella? sp.*
	9g	*Trigonia reesidei*

(Brackets at left indicate Range of pseudo-quadrate Trigoniae and Range of *Trigonia reesidei*; bracket at right indicates Range of *Dufrenoya*.)

Mural type. In this transitional zone similar accumulations of turritelloid gastropods form banks both in the brown and in the gray beds. A more precise contact between the Pedregosa member and the Mural limestone within the passage zone can be selected arbitrarily either on paleontological ground, at the lowermost gray limestone bed with *Orbitolina texana*, or lithologically, at the basal gray limestone bed of the Mural type. This, however, may not be the same horizon in all different sections.

GEOLOGICAL MAP AND PANORAMIC VIEW OF THE NINETY ONE HILLS AREA

The geological map of the Ninety One Hills area, in which is the type locality of the Lowell formation, covers the ground immediately north of the international border in the vicinity of the international monument No. 91, southeast of Bisbee Junction on the Southern Pacific Railway, and southwest of the Schlaudt Ranch. South-

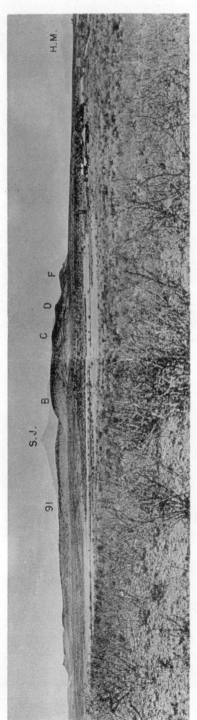

FIGURE 1. NINETY ONE HILLS AREA. WESTWARD VIEW.
Glance-Mural sequence. 91, International monument; S. J., San José Mts. and southern part of H. M., Huachuca Mts.,—in Mexico. Schlaudt Ranch and Ridge (basal Lowell) in the middle right. B-F, Mural limestone.

FIGURE 2. BLACK KNOB RIDGE. SOUTHWARD VIEW.
MO-L, Morita-Lowell contact. MU, Mural limestone.

LOWER CRETACEOUS SEQUENCE SOUTHEASTERN ARIZONA

FIGURE 1. MULE MOUNTAINS AT SHATTUCK DEN MINE. NORTHWARD VIEW.
Glance-Mural sequence. Four escarpments below Mural Hill form the Mural Hill Ridge. Mexican Canyon fault in the extreme right.

FIGURE 2. NINETY-ONE HILLS AREA. SOUTHWARD VIEW.
Glance and Morita in front. Light low ridges in the plain—Lowell formation. *Acanthohoplites* beds north of Hill E. B-F, Mural limestone.

LOWER CRETACEOUS SEQUENCE IN SOUTHEASTERN ARIZONA

west of the Schlaudt Ranch is the low Schlaudt Ridge, composed of the upper part of the Morita formation and the Pacheta member of the Lowell formation. This ridge is separated by a fault from a large, gently ascending general incline, which in the southeastern part is a structurally uninterrupted sequence of the Lowell formation from the top of the Morita formation to the base of the Mural limestone, but in the southwestern part of the mapped area, terminates in bold steep ridges of displaced Mural limestone. The Mural limestone forms Monument Ridge and Hill F which apparently once were continuous but at present are separated by a saddle This evidently resulted from a thrust which left the lower part of the Mural limestone in place in the middle of the original ridge but displaced its upper part to the northeast as an isolated block in which continuous Hills B, C, D, and the topographically lower Hill E are prominent (Pl. 3, fig. 1; Pl. 4, fig. 2; Pl. 27).

The standard section of the Lowell formation in the southeastern part of the ascending incline is not broken by faults (line A-A on Pl. 27). Although the northwestern part of the incline is faulted, all the members of the Lowell formation are mappable within the individual blocks, and the preservation of fossils in the strata is as good as in the southeastern part of the area, in certain units even better.

The principal fossiliferous strata can be located easily in the field by noting the member boundaries and outstanding beds on the map and in the panoramic picture, and by consulting the "Section A-A in the Ninety One Hills" and Tables 1 and 2 which show the vertical distribution of index fossils.

The panoramic view (Pl. 1) should be especially serviceable for visual orientation and quick locating of the stratigraphic and fossiliferous units along the section A-A of the map. The picture was taken from the top of Hill B, and its depth is about 6 miles. The international border passes from the upper third of the right-hand margin toward the middle of the upper margin of the picture approximately to the point where the black smoke cloud marks a train emerging from the tunnel on the Bisbee-Douglas railroad. The trend of the border can be located in the picture by the imaginary line along which all the strata pitch beneath the alluvial plain, with a draw in the distance on Mexican territory. The symbols of the panoramic view are the same as those of the map. The view shows all the strata of the Lowell formation from the Pacheta member, division 9g (with *Trigonia reesidei* and *Acila schencki*), to the Quajote member, division 3d (with *Acanthohoplites schucherti, Beudanticeras victoris*, and *Trigonia weaveri*). The stratigraphically higher units of the Lowell formation and the Mural limestone are on the right, beyond the limits of the picture.

METHODS IN THE FIELD AND IN THE LABORATORY

A glance at the panoramic view of the type area will suffice to visualize the nature of the field in which low ridges composed of harder strata alternate with depressions filled with alluvium and detritus. To have as complete a sequence as possible, the method of digging trenches across the strike of the strata and down through the alluvium to the bedrock was introduced with surprisingly gratifying results. The units rich in ammonites—Joserita member, division 8e; Cholla member, division 5g; and Quajote member, divisions 3d and 3b—were discovered by this procedure and later located at many points of the field. The buried condition of such strata probably explains the fact that no ammonites were reported from the Bisbee area by the previous workers.

The ammonites and lamellibranchs of the Pacheta member are thoroughly silicified, withstand weathering well, and are easily leached out of the rock with hydrochloric acid. However, the fossils from other strata, generally hard, solid rocks, had to be cleaned from the matrix by mechanical means in the laboratory. Good specimens naturally weathered out come only from a few horizons, such as the soft argillaceous beds of the Pedregosa member, in the upper part of the Lowell formation.

As the purpose of photographic illustrations of the fossils is to show pictorially the important characters which are discussed in the text, the method of illuminating all figured specimens from the same direction was not followed. This practice is satisfactory for the reproduction of homogeneous series of specimens but not for examples of radically different geometrical configurations, nor does it bring out advantageously the diagnostic features of a species. None of the photographic illustrations of the fossil specimens has been retouched. Many additional details may be seen with a magnifying glass.

MURAL LIMESTONE

The Mural limestone crosses the Bisbee area diagonally in a northwest direction, and from the heights of the Mule Mountains one can trace it with the naked eye and field glasses for a long distance across the international border into Mexico, in the ridges and hills all trending southeastward to an unknown distance. In the studied sections of the Bisbee area, the Mural limestone is composed of three principal units: a) the basal thinner-bedded limestone layers with *Orbitolina texana*, b) the massive "rudistid" limestone, and c) again the thinner-bedded limestone with *Orbitolina texana*. The contact between the Pedregosa member and the basal bedded limestone (a) is best seen in the section of Hill E of the Ninety One Hills (Pl. 4, fig. 2; Pl. 27). The uppermost beds with *Orbitolina texana* (c) have been located between the massive rudistid reefs (b) and the base of the Cintura formation on the east side of the Mule Mountains, north of the Bisbee-Douglas highway. At many localities with Cretaceous deposits in southeastern Arizona, where the Cintura formation has been removed by erosion, the upper strata with *Orb. texana* are also eroded, and the massive rudistid limestone forms the caprock of the Cretaceous sequence.

In the basal beds of the Mural limestone, small brachiopods, corals, specimens of *Lima muralensis*, and large forms of *Lunatia*? sp. often occur. The massive limestone is usually replete with *Radiolites*? sp., whereas the specimens of *Caprina* sp. (Pl. 11, figs. 4–5) are comparatively rare and come from the thinner-bedded layers below the reef.

INTERPRETATION OF PALEONTOLOGICAL STRATIGRAPHY

Under a separate heading "Boundary between the Aptian and the Albian" elsewhere in this paper are discussed: the standard Aptian-Albian ammonite zones accepted in the present research; the stratigraphic position of the revised and restricted parahoplitan and acanthohoplitan faunas; the correlation of the Arizona and Caucasian index zones with the standard ammonite succession; and the reason for placing the Aptian-Albian boundary between the *Immunitoceras nolani* and the *Hypacanthohoplites? jacobi* zones.

The Lowell formation contains three successive ammonite faunas: (1) the Parahoplitan fauna of the basal Pacheta member, (2) the Dufrenoyan fauna, which ranges from the middle of the Joserita member into the Cholla member, and (3) the Acanthohoplitan fauna of the Quajote member.

The faunas of the lower part of the Lowell formation, including the lamellibranchs, are unknown elsewhere in the United States, barring the possibility of the presence of certain undescribed Trigoniae in the Cretaceous strata of Texas which may be related to the Trigoniae of the *abrupta* and *excentrica* groups of Arizona. The dufrenoyan fauna of the Joserita and Cholla members is correlated with the *Dufrenoya* fauna of Texas (Travis Peak), but the acanthohoplitan fauna of the Quajote member is again new for this country.

Of other ammonites collected within the Aptian part of the Lowell formation, *Paracanthohoplites meridionalis* (lower part of the Joserita member) is related to a Caucasian species. *Beudanticeras victoris* (Quajote member) resembles *B. walleranti* (Jacob) from the Albian of France but is a different species with ornamented inner whorls;[4] the stratigraphic position of the Arizona species below the beds with the acanthohoplitan fauna, that is, immediately below the equivalent of the *nolani* zone, places it high in the Aptian. *Colombiceras? brumale* (Quajote member) may be related to certan South American forms which as yet have not been adequately studied.

In the collected assemblages of lamellibranchs from the Pacheta and Joserita members there are no species in common with the described species from the Cretaceous strata of Texas. Lamellibranch faunas of a comparable general aspect were not introduced before the deposition of the Saavedra member. However, there are still no common species, with a few possible exceptions: the Barata limestone of the Saavedra member has yielded an incomplete specimen of *Pecten* (No. 91258) which has certain features in common with *Pecten stantoni* Hill, and I have collected from the upper Trigonia limestone of the Cholla member (D.6a) a specimen (No. 91838) which seems to be nearly identical with the illustrated internal mold of *Homomya? solida* Cragin (1893, p. 191, Pl. 39, figs. 3–4), a plaster cast of which I have through the courtesy of Dr. H. B. Stenzel of the Bureau of Economic Geology of Texas. Unfortunately both the types and the compared forms are very inadequately preserved.

The Trigoniae of the pseudo-quadratae and the *v-scripta–vau* groups of the Saavedra and Cholla members are closer to the Jurassic Trigoniae from the Malone strata than to the Texan Cretaceous Trigoniae. However, in the Barata limestone (D.7n) the Trigoniae of the *aliformis* group, introduced simultaneously with the pseudo-quadrate Trigoniae, give the entire paleontological assemblage a distinctly younger aspect.

The basal (Lancha) limestone and especially the topmost (Pedregosa) strata of the Lowell formation contain numerous high-spired turritelloid gastropods which at first glance direct the student to the group of "*Mesalia seriatim-granulata.*" The

---

[4] *Beudanticeras hatchetense*, a species described by Gayle Scott (1940, p. 1000, Pl. 56, figs. 3–5) from New Mexico, does not seem to be closely related.

ornamentation of such forms varies considerably in both horizontal and vertical directions, but my large material does not include specimens that can be readily identified with Roemer's species of Texas.

On the basis of the presented observations it may be inferred that the Aptian part of the Lowell formation holds no index fossils common to Texas below the *Dufrenoya justinae* zone (Cholla member, D.5g). Above this zone, but also above the strata with the Acanthohoplitinae, are two valid index species common to both regions: *Trigonia mearnsi* and *T. stolleyi* (Perilla member, D.2b). *T. mearnsi* is of special stratigraphic interest; in the Quitman Mountains of southwestern Texas it occurs with *Trinitoceras* above the zone of *Sonneratia trinitensis* but below the *Orbitolina texana* zone. *T. stolleyi* occurs in the Glen Rose formation. The Perilla member also yielded *Neithea vicinalis*, which is related to certain Texan species of a younger age.

Taken in its entirety the determinative fauna of the Lowell formation has exotic relations. Its ammonite fauna is, with the exception of *Dufrenoya*, almost totally foreign. The following list shows regions with comparable forms.

Acila (Truncacila) schencki	England, northwestern continental Europe, northwestern Africa, Venezuela, California.
Acanthohoplites berkeyi	France, Mangyshlak.
A. erraticus	Mangyshlak, Caucasus.
A. hesper	France.
A. impetrabilis	Mangyshlak, Caucasus.
A. schucherti	Caucasus, France.
A. teres	Mangyshlak.
Astarte adkinsi	France, England.
Beudanticeras victoris	France.
Columbiceras? brumale	South America?
Deshayesites? butleri	Undetermined.
Dufrenoya joserita	Texas.
D. justinae	Texas.
Exogyra lancha	South America, France.
Gervillia cholla	Western Europe.
G. heinemani	Western Europe.
G. rasori	Western Europe, South America.
Idonearca stephensoni	South America.
Immunitoceras immunitum	Mangyshlak, Caucasus, France.
Kazanskyella arizonica	Caucasus, Persia?
Lima cholla	France, Texas.
Neithea vicinalis	Texas, Western Europe.
Ostrea edwilsoni	Portugal, France.
Paracanthohoplites meridionalis	Caucasus, Mangyshlak?
Pecten (Chlamys) thompsoni	England, France?
Pteria peregrina	France.
Sinzowiella spathi	Mangyshlak.
Trigonia cragini	Texas (Jurassic).
T. guildi	Texas (Jurassic).
T. kitchini	Autochthonous.
T. mearnsi	Texas.
T. reesidei	South America, Texas?
T. resoluta	Texas (Jurassic).

T. saavedra............................Probably autochthonous.
T. stolleyi..............................Texas.
T. weaveri..............................Argentina, Texas?

Considering the data presented above, the stratigraphic sequence of the Bisbee group is interpreted as follows:

In southeastern Arizona the initial encroachment of the Cretaceous sea upon the pre-Cretaceous land mass, which is represented by the Glance conglomerate, cannot be dated exactly because of the lack of fossils in the overlying Morita formation. Nevertheless, there is no sufficient reason to accept Burckhardt's (1930, Pl. 12) contention that the Morita formation is of Hauterivian-Barremian age. Burckhardt correlated the Morita formation with the Las Vigas beds on the Rio Grande. Adkins (1932, p. 291), however, pointed out that the basal red sandstones near Indian Hot Springs, Texas, which are an apparent continuation of the Las Vigas beds, are successively overlain by strata with: (a) *Dufrenoya* (Upper Aptian, Gargasian), (b) *Douvilleiceras* (Albian), and (c) *Orbitolina* (Albian, Glen Rose). On the evidence of the stratigraphic succession in the Lowell formation and its paleontologic content, the *Parahoplites melchioris* zone is now inserted between the Morita formation and the beds with *Dufrenoya* (*see* Table 6). Since there is no widespread depositional discontinuity or a time interval between the Morita and the Lowell formations, save for the change of the facies, it is reasonable to assume that the Morita formation is not older than the Bedoulian (*Deshayesites deshayesi* time).

The parahoplitan and the dufrenoyan beds of the Lowell formation (Pacheta-Cholla members) belong in the Gargasian. The upper part of the Quajote member, possibly the entire member, is an equivalent of the Clansayan which I prefer to place in the Aptian for the reason presented elsewhere in this paper.

The Albian ammonite faunas of the *jacobi* and *trinitensis* zones have not been located in the Lowell formation or elsewhere in Arizona. However, the *Trigonia mearnsi* zone of the Perilla member (D.2b) is the obvious equivalent of the beds with *T. mearnsi* in the Quitman Mountains, Texas, which are between the *Sonneratia trinitensis* zone and *Douvilleiceras mammillatum* zone, the latter being definitely below the basal strata with *Orbitolina texana*.

The Mural limestone—the Arizona equivalent of the Glen Rose formation—begins with the *Orbitolina texana* beds which are overlain by the rudistid reefs, as in the marginal facies of the Glen Rose in Texas (Adkins, 1932, p. 316).

TWO SECTIONS IN THE BISBEE AREA

Two examined sections across the Mule Mountains, one to the north and the other to the northeast of the Ninety One Hills, are briefly described here to show the observed lateral variations of the Lowell formation within the Bisbee area.

MURAL HILL RIDGE SECTION

The Mural Hill Ridge begins at Winwood Addition, southeast of Bisbee, and culminates in Mural Hill—a prominent landmark of the Bisbee area and a summit in the Mule Mountains northwest of the Mexican Canyon fault (Pl. 2, 3; Pl. 5). The strata are in the northeastern flank of the Bisbee anticline. The ridge has four prominent escarpements caused by the presence of relatively harder rocks, all below Mural Hill, which forms the fifth escarpment (Pl. 4, fig. 1; Pl. 6, fig. 2).

No attempt was made to measure the entire sequence in this locality. Approximate thicknesses were estimated only for the exposed strata not covered by debris or talus.

The basal Glance conglomerate is not thick in this part of the Bisbee area. Its contact with the pre-Cambrian strata passes through Winwood Addition. The Morita formation is predominantly composed of maroon mudstones and siltstones which alternate with light-colored and brown quartzites and, less frequently, with layers of conglomerate and sandstone. It should be noted that the hard speckled quartzites, so characteristic of the Morita formation in the Ninety One Hills area and in the Black Knob Ridge, are predominant here in the upper part of the formation.

The *first escarpment* is a sheer wall of typical hard Morita quartzite, between 30 and 50 feet thick. The overlying Morita mudstone occupies the entire upward slope to the second escarpment.

The *second escarpment* is also composed of harder rocks: a distinctly cross-bedded brownish-buff sandstone and a somewhat lighter-colored quartzite above. The first limestone bed in the sequence rests directly on this quartzite and may be tentatively correlated with the basal limestone of the Lowell formation in the Ninety One Hills area.

This unfossiliferous yellowish-gray limestone (1) contains inclusions of the Morita mudstone. The examined contact indeed resembles the Morita-Lancha boundary in the international border area, but, unlike it, here are observed not only the sudden replacements of mudstone and shale by limestone, which is the case with the contact in the type locality, but also angular chunks and fragments of the Morita rocks actually entrapped in the limestone. However, there is a difference in succession. Instead of the very fossiliferous basal limestones of the Ninety One Hills area, the unfossiliferous basal limestone is succeeded here by a maroon mudstone and shale, another yellowish-gray limestone, a bed of quartzite, and a very characteristic gray fragmental limestone, all without fosssils. Higher up in the section is a 1-foot bed of dolomite (2), white limestone beds about 10 feet in aggregate thickness (3), and a maroon mudstone 4 feet thick (4). There is another white limestone (5) which rests on the preceding unit. This limestone weathers round, is 7 feet thick, and is overlain by 12 feet of shale (6). Above is a whitish-green pitted quartzite 6 feet thick (7); its lower part is a solid ledge, but it grades upward into thinner layers. The following massive brownish-maroon quartzite (8) is at least 15 feet thick. The next unit—a maroon mudstone (9) with a hard-bedded quartzite near the top and some shale above—has an aggregate thickness of 10 feet. Still higher are prominent ledges of quartzite (10), brownish-white, but changing to greenish toward the top, 10 feet thick. A maroon sandstone (11) above it forms a slope and attains a thickness of 16 feet. Next, a prominent ledge-forming brownish-buff quartzite (12), with some shale in partings, is a very hard rock, very resistant to weathering and a good key bed throughout the area. Although not represented as a prominent escarpment, and placed in the slope in the examined sections, in the next ridge to the north it forms a hill with a cuesta (Pl. 4, fig. 1; Pl. 6, fig. 2). A little higher in the section and down the dip is a greenish dolomite (13) 3 feet thick. Here begins the upward slope that leads toward the third escarpment.

The first stratum to attract attention in this part of the section is a gray, yellow-spotted limestone (14), 8 feet thick, with oysters and poorly preserved scabroid Trigoniae, the first fossiliferous stratum encountered in the described section. In the Ninety One Hills area the range of scabroid Trigoniae is from the Saavedra member to the Perilla member. If this limestone represents one of the lower fossiliferous beds of the Saavedra member, of which there is no assurance, then the overlying 25 feet of buff ledge-forming sandstone (15) interbedded with quartzite in the upper part may be an equivalent of the Chapparal sandstone of the type area. This sandstone is separated by a gray limestone (16) from the 75 feet of stratified hard quartzites (17) which form nearly impassable cliffs and which compose the base of the *third escarpment*. These quartzites have no direct counterpart in the Ninety One Hills.

On the third escarpment is a 1-foot bed of gray fossiliferous limestone (18). The overlying 7 feet is buff quartzite (19) which composes the hill of 6320 feet elevation. This quartzite is thin-bedded and underlies a suite of predominantly gray and black limestone beds which form the next saddle and the upward slope that leads toward the fourth escarpment. I regard these strata as the transitional zone from the Pedregosa member of the Lowell formation to the basal beds of the Mural limestone.

The first prominent ledge in this series, a gray limestone (20), rests on a weaker limestone, weathers round, and is 6 feet thick. The overlying shaly limestone layers (21) with *Lunatia*? sp., scabroid Trigoniae, and oysters are 10 feet thick and separated by a prominent ledge of harder limestone (22) from the unit of predominantly argillaceous black fossiliferous (*Lunatia*?) limestone beds (23) with some shale near the top which occupy the deepest part of the saddle between the heads of two gullies. In the lower part of the upward slope toward the fourth escarpment a gray ledge-forming limestone (24) a few feet thick contains shells of very large ammonites. *Douvilleiceras? muralense*, sp. nov. (Pl. 26, figs. 1–2) was collected from this stratum. Seventy feet of gray and buff thin-bedded sandstones (25) forms the principal part of the fourth escarpment. Still in the slope is the drab-olive gastropod (small turritelloid shells) limestone (26) 2 feet thick. The hard thin-bedded sandstone (27) forms the top of the *fourth escarpment*.

Above the fourth escarpment, in the saddle between the two gullies nearest Mural Hill, is a gray soft sandstone (28) about 25 feet thick. Upward it terminates in a gray ledge-forming 1-foot limestone (29) with inclusions of quartzite and other rocks.

The lower part of the slope leading to the sheer cliff of Mural limestone at the top of Mural Hill is composed of gray and buff sandstone (30) 25 feet thick. Above is 15 feet of gray limestone beds (31) made up of individual ledges each 2–3 feet thick. This limestone contains large turritelloid gastropods.

Between the uppermost ledge of the last unit and the vertical cliff of the massive Mural limestone is talus, but through the talus a part of the slope is seen to be composed of beds of brown sandstone, gray arenaceous shale, and ledges of quartzite (32). The gray shale is underlain and overlain by layers similar to the olive gastropod limestone (26). The vertical wall of the Mural limestone is 50 feet thick. Above it, on the northeast slope of Mural Hill, are thinner-bedded limestones with *Orbitolina texana*.

It is possible that the entire sequence of strata between the Glance conglomerate and the Mural limestone, as it is known in the Ninety One Hills area, is also represented in the Mural Hill Ridge. Yet the lithologic difference is considerable, especially in the lower part of the section. Many of the principal stratigraphic units of the Lowell formation are not recognizable here, and no index fossils of that formation have been found. Compared with the type locality, the Mural Hill Ridge section shows: (1) a greater upward development of the maroon shales of the Morita type *above* the basal limestone, (2) a greater development of quartzites, (3) brown and yellow limestones (Pedregosa type) alternating with light-gray limestones (Mural type) in the upper part of the section, which makes it rather difficult to establish the base of the Mural limestone in this area.

Lateral lithologic variations must be considered, of course, as the actual depositional areas were much more distant than they appear in the anticline. However, the general consensus regarding the geology of the Bisbee area is that the abundance of thick quartzites and silicification of the strata may have resulted from metamorphism. The principal area of mineralization and the main ore bodies of Bisbee are close. Since the only larger intrusive body that might be considered responsible in this connection—the near-by stock of granite porphyry—was regarded by Ransome (1904, p. 78; Pl. 1) as pre-Cretaceous, a postulate of its post-Cretaceous age would be necessary to account for such lithologic changes. It is true, that Ransome also observed small dikes of porphyritic rocks, probably a monzonite porphyry, which cut across the Cretaceous strata, but these were considered as independent and younger than the stock (Ransome, 1904, p. 84). These dikes have not been studied, however, and their source or relation to the granite porphyry is unknown.

BLACK KNOB RIDGE SECTION

Black Knob, a prominent landmark in the Bisbee quadrangle, is in the extreme southeastern termination of the Mule Mountains (Ransome, 1904, Pl. 1). On the topographic and geologic maps published by the Arizona Bureau of Mines it is erroneously shown in the small hills composed of Mural limestone which are north of Paul Spur and about 5 miles southeast of the true location. The number "2" on Plate 2 indicates the correct position of Black Knob.

In Black Knob Ridge probably the entire equivalent of the Lowell formation is present between the Morita formation and the Mural limestone. However, the fossils are satisfactorily preserved only in the basal and in the upper beds of the sequence. The strata are in the northeastern and eastern flank of the curved Bisbee anticline. Unlike the section of the Mural Hill Ridge, here the contact between the topmost strata of the Morita formation and the basal beds of the Lowell formation is at the summit of Black Knob, whereas the Mural limestone is in the foothills of the ridge (Pl. 3, fig. 2). There are displacements within the section, and the thicknesses given here are only approximate.

Stratigraphically above the topmost blue and maroon shales of the Morita forma-

FIGURE 1. PART OF THE TYPE SECTION A-A' IN NINETY ONE HILLS. LOOKING WEST.
Espinal grit overlies dolomite with *Dufrenoya joserita*. Lenses with *Paracanthohoplites* in front under detritus. Hill B in the background.

FIGURE 2. MURAL HILL RIDGE FROM WINWOOD ADDITION.
Four escarpments of the ridge. Glance and Morita in front. Lowell in the middle.

FIGURE 3. SECTION AT LEE SIDING.
Pedregosa member and Mural limestone. Volcanics in the background.

LOWER CRETACEOUS SEQUENCE IN SOUTHEASTERN ARIZONA

FIGURE 1. PART OF THE TYPE SECTION A-A' IN NINETY ONE HILLS. LOOKING EAST. Upper limestone with *Trigonia crayini*. Sandstone with *Araucarioxylon* and limestone with *Dufrenoya justinae* in front.

FIGURE 2. GLANCE CONGLOMERATE AT WARREN. Arched Glance conglomerate in the crest of Bisbee anticline.

LOWER CRETACEOUS SEQUENCE IN SOUTHEASTERN ARIZONA

tion are the basal limestone beds of the Lowell formation with *Trigonia reesidei* and *Kazanskyella arizonica* (1). These beds are overlain by 23 feet of maroon shale, thin-bedded limestone, cross-bedded sandstone, in places conglomeratic, with silicified wood and vertebrate remains, and thin-bedded reddish calcareous sandstone from which specimens of a small simple-ribbed *Pecten* (*Chlamys*) sp. were collected (2). The overlying Black Knob dolomite (3) is 56 feet thick, exceeding that in Ninety One Hills area. But the Black Knob quartzite (4) is about 10 times thicker, 412 feet, than in that area. It is composed of buff and white quartzite beds which grade upward into quartzitic sandstone.

The next unit (5), 282 feet thick, is made up of sandstone which in the middle alternates and is replaced by limestone beds with pseudo-quadrate Trigoniae. A sandstone with gastropods is at the top of this unit. Since, in the type locality of the Lowell formation, the pseudo-quadrate Trigoniae do not occur below the Saavedra member, the 412 feet of the underlying quartzite (4) may represent not only the Black Knob quartzite of the international border area but also the entire Joserita member.

Unit (6) is composed essentially of grayish-white shale with buff sandstone in the middle part. The entire unit is 274 feet thick. The overlying unit (7), 262 feet thick, is composed of brown-buff argillaceous sandstone with silicified plant remains, ripple marks, and concretions, which, upwards, alternates with, and higher up is replaced by, quartzites containing numerous large silicified trunks of *Araucarioxylon*, some of which are normal to the bedding planes.

In the next unit (8) buff quartzites alternate with layers of sandstone containing plant remains and reptilian bones. Upward the clastic rocks are replaced by gray limestone with scabroid Trigoniae. This unit is 218 feet thick.

Unit (9), 111 feet thick, is composed of gray shale with beds of grayish limestone, 1–2 feet thick. At the top there is maroon shale and gray limestone with numerous specimens of *Trigonia mearnsi*. This limestone, already in the foothills of the Black Knob Ridge, corresponds with division 2b of the Pedregosa member of the Lowell formation in the international border area. Above is grayish-buff quartzite (10) 42 feet thick. Within the overlying buff sandstone (11), 142 feet thick, are gray limestone beds, up to 10–15 feet of individual thickness, with lamellibranchs and gastropods. At the top of this unit is 50 feet of drab-gray limestone weathering buff with thin layers of gray limestone in which are abundant specimens of poorly preserved echinoids and ammonites. The sequence terminates in the Mural limestone.

EASTWARD SECTIONS

A few isolated areas with strata presumably of Cretaceous age are known northeast of Bisbee and the Mule Mountains. In the Dragoon Mountains (Darton, 1925, p. 142), in the low hills northwest of Cochise, and along the foothills in the terminal part of the Dos Cabezas Mountains, the post-Permian, partly homogeneous and calcareous, conglomerate crops out, overlain in places by fossiliferous limestones and red shales. These strata have not been studied adequately.

East of the Bisbee area the Cretaceous sequence has been briefly examined: (1)

in the isolated hills located on the highway between Bisbee and Douglas southeast of the Mule Mountains, northeast of the Hay Flat, and immediately north of Paul Spur (Pl. 2); (2) at Lee Siding between the Pedregosa and Perilla Mountains (Pl. 2, 4; Pl. 6, fig. 3); and (3) in the Guadalupe Mountains in the extreme southeastern corner of Arizona (Pl. 2, 5). In the eastern direction the lower part of the Lowell formation either is not exposed, as in the first two localities, or is represented by strata entirely different from those of the type area, as in the Guadalupe Mountains.

Thus in the hills (1) the massive rudistid (*Radiolites*? sp.) reef of the Mural limestone rests on the thinner beds of gray limestone with *Orbitolina texana*, giant shells of *Lunatia*? sp., *Caprina* sp. (Pl. 11, figs. 4–5), *Lima muralensis* sp. nov. (the holotype collected at the base of the Mural limestone in the Ninety One Hills area is illustrated in Fig. 2 of Pl. 9), *Pyrina* sp., and numerous specimens of *Enallaster* sp. All these strata belong in the Mural limestone.

At Lee Siding (2) the Mural limestone at the top of the section is composed of massive rudistid reef and underlying thinner limestone with *Orbitolina texana*. Below are yellow limestones and shales of the Pedregosa member with lamellibranchs (*Clementia* sp.,[5] *Cyprimeria* sp., *Protocardia* sp., *Lunatia*? sp.) and turritelloid gastropods. In the lower part of the section are limestones, replete with unidentifiable fragments of lamellibranchs, and quartzites. These exposed basal strata most probably take the place of the Perilla member of the type area.

Much more of the sequence is seen in the two sections examined in the Guadalupe Moutains (3) where the Cretaceous strata, dipping east, are nearly encircled by the post-Cretaceous volcanics. One of these cross sections has been taken at the northern termination of the exposed strata south of McDonald Ranch near Sycamore Creek, and the other in the Guadalupe Canyon $2\frac{1}{2}$ miles north of Hall Ranch (Baker Springs), closer to the international border. The Cintura formation is absent in this area, and the sequence terminates in the Mural limestone.

In the section near Sycamore Creek the oldest exposed stratigraphic units are observed in the floor of the desert, in the low hills, and in the low ridges west of the Guadalupe Mountains proper. 1. The apparent basal unit is formed of thin-bedded, bluish-gray limestone with numerous small fragments of lamellibranchs. The stratigraphically higher blue limestones and dolomites are still in the desert floor. 2. The first foothills are made up of hard, cross-bedded and streaked, gray and brown sandstones capped by their weaker varieties which form a topographic depression below the next unit. 3. Here begins the uniform ascending slope which terminates in the Mural limestone. This unit is composed of soft yellow, gray, and drab sandstones which become harder toward the top. Poorly preserved specimens of pseudoquadrate and scabroid Trigoniae, and fragmentary specimens of ammonites have been collected from these beds. 4. In the next unit darker sandstones are superimposed first by thin-bedded yellow limestones and dolomites, and in the upper part by limestone beds with pecten and oyster banks. Toward the top the individual beds thicken, and the number of the banks increase. The unit terminates in thick ledges of gray and brown limestone with lamellibranchs. 5. In this unit calcareous

---

[5] This form seems to be close to *Clementia* (*Flaventia*) *ricordeana* (d'Orbigny) from the Lower Greensand, Atherfield, England (*see* Woods, 1913, p. 189, Pl. 29, figs. 18a–b; Woodring, 1927, p. 28).

grits and green sandstones with some shale are overlain by yellow limestone which upward alternates with ledges of much harder fossiliferous gray limestone (numerous specimens of *Epiaster* sp. in the uppermost ledge). 6. Ledges of hard cross-bedded limestone separate unit 5 from the prominent massive member of the Mural limestone.

The upper part of the sequence is better exposed to the south, in the Guadalupe Canyon, where the Perilla and Pedregosa members of the Lowell formation, the *Orbitolina* beds, the massive limestone of the Mural, and their stratigraphic contacts are as well developed as in the type area, whereas the preservation of fossils is even better. The yellow shales of the Perilla with *Trigonia mearnsi*,[6] *T. stolleyi*, and *Astarte adkinsi* are 17 feet thick. The overlying 148 feet of thin-bedded yellow shales and sandstones is less fossiliferous. Within the upper 18 feet of these strata conspicuous cross-bedded sandstone and hard brown limestone separate the Perilla from the Pedregosa. The first 50 feet of the latter member also are yellow and brown shales and sandstones. A hard brown limestone separates these strata from the upper division in which yellow and gray shales alternate and which culminates in dark-gray limestone beds with turritelloid gastropods. Above is 65 feet of thin-bedded gray limestone with *Orbitolina texana*. At the top of the sequence is the massive rudistid limestone of the Mural.

WESTWARD SECTIONS

West of the type area Lower Cretaceous strata are present: (1) south of the town of Tombstone, which is northwest of the Mule Mountains (Pl. 2, 9); (2) in the southwestern side of the Huachuca Mountains about 30 miles west of Bisbee (Pl. 2, 6); and (3) in the Patagonia Mountains approximately 50 miles south of Tucson and 6 miles north of the international border (Pl. 2, 7).

The stratigraphic relation of the Cretaceous outlier south of Tombstone (1) has been discussed by B. S. Butler, E. D. Wilson, and C. A. Rasor (1938, p. 22) as follows:

"The Bisbee group of the Mule Mountains cannot be directly traced to the Tombstone district. The gap between the hills east of Sandy Bob Ranch where the Glance conglomerate passes below the alluvium at the most northerly point to which it can be traced from Bisbee and the low hills north of Government Draw where similar conglomerate appears is about 6 miles. There are many small outcrops of similar conglomerate between Government Draw and Tombstone. James Gilluly (written communication) who has mapped this area believes that the conglomerate at the base of the Mesozoic section at Tombstone is the equivalent of the Glance, though it may be slightly younger inasmuch as *the Comanche sea evidently overlapped this region from the south* [Italics mine].

" 'The Mural limestone thins northward from Bisbee as far as it can be traced in the Mule Mountains, and it seems likely that it pinched out entirely, and the Cintura may have directly overlain the Morita with a few more miles to the north. The Cintura resembles the Morita very closely in lithology, and were it not for the intervening Mural limestone it is unlikely that the two formations would have been distinguished at Bisbee. The Mesozoic section above the basal conglomerate at Tombstone consists of rocks *very like those of the Morita and Cintura* [Italics mine], though the proportion of the several varieteis cannot be closely matched.' "

The paleontological collections from Tombstone are poorly preserved lamellibranchs that cannot be compared with assurance to any of the lamellibranch assemblages known from the Bisbee area. In my opinion all the geological evidence pre_

---

[6] This is the place where the specimens of *T. mearnsi*, now on deposit in the U. S. National Museum, were collected by E. A. Mearns in 1892. (*Compare* Ransome, 1904, p. 70, footnote).

sented indicates a post-Mural, and possibly even post-Cintura sea incursion from the south.

The section examined in the Huachuca Mountains (2) is in the upper part of the Parker Canyon above the road between the Montezuma Pass (north of the international border) and Canille. In the canyon the strata are uniformly vertical. A little above the meadow on which the old C. C. C. camp was located are exposed impure limestones with most typical examples of *Ostrea ragsdalei* Hill and numerous specimens of *Glauconia* aff. *G. branneri* (Hill).[7] Higher up in the canyon but down the section, the predominantly buff sandstone beds are separated by a rhyolite (?) from the thin-bedded reddish slates composed of consolidated feldspathic material. These strata persist nearly to the top of the Huachuca Mountains which in this section consists of an almost horizontal snow-white limestone-breccia (900–2000 feet estimated thickness). According to Mr. Carl Alexis (personal communication) this breccia, or conglomerate, stratigraphically rests on the Permian limestone and, in a flexure cut by cross-fault, underlies the Cretaceous strata.

Strata with *Ostrea ragsdalei* have not been recorded elsewhere in Arizona. This unit undoubtedly belongs in the Glen Rose formation. Unfortunately its relation to the Mural limestone is unknown.[8] Imlay (1944, p. 1037) stated that beds with *Exogyra arietina* are present "on the summit of the next to highest peak in the Huachuca Mountains" which, so defined, is Carr Peak north of the Miller Canyon, a locality not yet adequately examined.

In a preliminary report (Stoyanow, 1937, p. 296) I designated the Cretaceous sequence in the Patagonia Mountains (3), with a thickness variously estimated between 4500 and 5500 feet depending on the selected sections, as the Patagonia group, and the formation containing limestone beds with an abundant *Stoliczkaia* fauna, in the upper part of the group, as the Molly Gibson formation. For more accurate measurements and lithological observations I am obliged to Mr. M. B. Lovelace.

1. At the base of the sequence thin and papery red shales rest on the pre-Cambrian schist and granite, and locally probably also on the Paleozoic strata. These shales are capped by a thin layer of andesitic agglomerate.[9] Above is 100 feet of hard reddish sandstone in which lenses of conglomerate, sometimes just larger pebbles scattered in a finer sandstone, and thin layers of quartzite occur at irregular intervals. This unit grades upward into finer-grained and evenly bedded arenaceous shales which are coarser and contain larger angular fragments in their upper layers. This unit is about 700 feet thick. 2. Higher in the sequence are thin beds of andesitic flows and associated agglomerates, mainly purplish and grayish, about 250 feet in aggregate thickness. 3. Above is a monotonous series, 3400 feet thick, of brown gray, grayish-white, and red argillaceous shales in layers up to 6 inches thick. These shales are badly folded, faulted, and crumpled, with rapidly changing attitudes. 4. The 1000 feet of strata of the next unit, the Molly Gibson formation, is composed mainly, and especially in the lower part, of brown, gray, chocolate, and red argil-

---

[7] This variety is closer to a form from the Glen Rose of Texas (Dane, 1929, Pl. 4, fig. 10) than to Hill's types from Arkansas.

[8] In a preliminary report (Stoyanow, 1937, p. 296) the lower series was mistaken for an equivalent of the Morita formation.

[9] Some geologists working in this area regard this and higher volcanics as essentially rhyolitic.

laceous and siliceous shales. In the upper part of the formation are first, bluish limestone beds; then, thin and soft varicolored arenaceous and argillaceous shales; and eventually soft and crumbly, light-gray and pinkish, arenaceous lamellibranch limestone beds with abundant and new forms of *Stoliczkaia*, some of which are described in the paleontological part of this paper. A specimen of *Pentagonaster?* sp., with numerous marginal plates, also was collected from these strata. Near the Molly Gibson Mine this formation is in falut contact with the Paleozoic strata (Cambrian-Permian), of which the American Peak is composed. Undoubtedly this is the locality from which Schrader (1915, p. 53) reported a fragment of an ammonite which was tentatively referred to *Acanthoceras*. 5. Stratigraphically higher, a series of dark-blue and black hornstones, which alternate at intervals with siliceous and papery argillaceous shales, is 1500 feet thick.

The total thickness of the Patagonia group, as measured by Mr. Lovelace, is about 5850 feet. The Molly Gibson formation must be about the same age as the Grayson (Del Rio) formation of Texas (*Stoliczkaia dispar* zone). The Patagonia group may well represent the entire sequence from the Fredericksburg to the Buda. It may be argued that the deposition of the lower part of the Patagonia group might have been co-existent with that of the Cintura formation of the Bisbee group, and that the difference in the lithologic composition could be ascribed to the presence of two more or less isolated depositional areas. However, no evidence of volcanic activity has been recorded for any member of the Bisbee group. At present it cannot be said whether there were considerable interruptions in deposition of the Patagonia group, or whether the lower part of the group embraces more than Washita time.

BOUNDARY BETWEEN THE APTIAN AND THE ALBIAN

The so-called Clansayes beds and their equivalents occupy an important place in the ammonite zonal schemes of the Lower Cretaceous. Jacob (1905, p. 430) originally interpreted the Clansayes ammonite faunas as intermediate between those of the *Dufrenoya "furcata"* and *Douvilleiceras mammillatum* zones. Stolley (1908a, p. 242–247) correlated the Clansayes with the *nolani* clay of Bettmar and Sarstedt, yet he mistakenly believed that the parahoplitan faunas were *younger* than *Immunitoceras nolani* and criticized Anthula for placing the strata with *Parahoplites* of the Caucasus in the Aptian (Stolley, 1908b, p. 321–325). As a matter of fact, Stolley's "Parahopliten Schichten" of northern Germany, with *Hypacanthohoplites? jacobi*,[10] which rest on the *nolani* clay and underlie the strata with the *Leymeriella tardefurcata* fauna, do not contain true parahoplitan ammonites. The error was caused by the failure to discriminate between the parahoplitan fauna proper and the acanthohoplitan fauna. This was accomplished by Sinzow (1908) whose work apparently remained unknown to Stolley at that time. The stratigraphically lower *melchioris* zone does not seem to be typically developed in northern Germany. Whether or not any ammonites of the *jacobi* zone are congeneric with *Acanthohoplites* is open to discussion.

The change from the parahoplitan and acanthohoplitan (Aptian) to the leymeriellan and hoplitan (Albian) succession is great. No species of the Parahoplitinae and

---

[10] Spelling of generic name changed for the same reason as in case of *Acanthoplites*.

Acanthohoplites, as defined in this paper, has the venter either strongly depressed or with interrupted ventral ribbing in the adult stage of the growth. On the other hand, this interruption is an important morphological feature in many Albian forms. The Aptian Cheloniceratidae are certainly quite apart; Spath (1923a, p. 35) tentatively considered this family as derived from the Desmoceratidae, and Roman (1938, p. 426. *Compare* Spath, 1923a, p. 64) from the Parahoplitidae. Roman's opinion seems more plausible because of the definite similarity in the shell form between the early development of *Acanthohoplites* and the advanced stages of *Cheloniceras*. *Cheloniceras* and "*Diadochoceras*" still await an ontogenic study of the shell and the suture.[11] The Cheloniceratidae are less widespread geographically than the Parahoplitidae. The described American forms referable to the Cheloniceratidae are doubtfully related to the European species and are still insufficiently known to be interpreted by intercontinental stratigraphy and paleogeography. To date, in the United States east of California, *Cheloniceras* is represented only by one specimen in Texas (Scott, 1940, p. 1005). The Cheloniceratidae are apparently present in the Rio Nazas area, Durango, north-central Mexico. The Rio Nazas fauna (Burckhardt, 1925), however, needs thorough revision.

It should also be noted that the acanthohoplitan and cheloniceran faunas do not pass into the *Leymeriella tardefurcata* zone—i.e., the classical base of the Albian. In his early essay Jacob (1905, p. 431) emphasized that the Clansayes fauna, notwithstanding its "transitional" nature, is pre-eminently characterized by *Immunitoceras nolani*, *Acanthohoplites bigoureti*, and "*Diadochoceras*" *nodosocostatum*. Moreover, it is doubtful that true species of *Acanthohoplites* are present in Stolley's "Parahopliten Schichten" between the *nolani* clay and the "Tardefurcata Schichten" of northern Germany—that is, in the typical *jacobi* zone of Spath (1923a, p. 4).

Unfortunately, the status of "*Parahoplites*"-"*Hypacanthohoplites*"-"*Acanthohoplites*" *jacobi* is considerably confused. Originally Spath (1923a, p. 64) selected "*Acanthoceras*" *milletianum* (d'Orbigny) var. *plesiotypica* Fritel="*Parahoplites*" *jacobi* Collet (1907, Pl. 8, fig. 1 only) as the genotype of his then new genus *Hypacanthohoplites*. In 1939 he transferred Fritel's variety, together with the abovementioned specimen of Collet, to "*Acanthohoplites*" *plesiotypicus* and designated another specimen, illustrated by Collet in his Figure 3, Plate 8, as "*Acanthohoplites*" *jacobi* (Spath, 1939a, p. 236–239). In the same paper the genus *Hypacanthohoplites* was apparently reserved for *Ammonites milletianus* d'Orbigny.

The reason why I question the presence of the typical representatives of *Acanthohoplites* in the *jacobi* zone is as follows. The tuberculate forms identified by Collet and Stolley as "*Parahoplites aschiltaensis*" and "*P. uhligi*" may at best be placed tentatively under the subfamily Acanthohoplitinae, and the latter form, with strongly developed umbilical tubercles and with trifurcating ribs in the adult stage (Collet, 1907, p. 523, Pl. 8, figs. 7–8), is certainly not conspecific with *Immunitoceras? uhligi* (Anthula, 1899, Pl. 10, figs. 1a–1b), as already noted by Spath (1939a, p. 238). Collet compared this form, probably not without good reason (*compare* Spath, 1939a), with "*Acanthoceras*" *milletianum* (d'Orbigny) var. *elegans* Fritel (1906, p.

---

[11] The difference between the sutures of these two genera was noticed by Sinzow (1906, p. 177) in the description of his species "*Douvilleiceras*" *subnodosocostatum*.

246, fig. 3). Its first lateral lobe (Collet, 1907, p. 527, fig. 7) is indeed too asymmetrical for *Acanthohoplites*.

Further, of the two text-figures which illustrate *"Parahoplites" jacobi*, Collet's Fig. 1, p. 520, shows the whorl section. Presumably this sketch was made from one of the three specimens photographed in his Plate 8. Although I can not estimate accurately the dimensions from a bibliofilm in my possession, the figured specimen undoubtedly has a depressed ribbing on the venter in the advanced or adult stage. No true species of *Acanthohoplites* has depressed ventral ribs except in the earliest whorls.[12] Collet's Figure 2, p. 521, shows a suture with more symmetrical saddles than in d'Orbigny's (1840, Pl. 77, fig. 3) drawing of the suture in *Ammonites milletianus*. However, the contracted external lobe and the narrow first lateral saddle are not typically acanthohoplitan.

In the absence of a revised paleontological study of the ammonites from the *jacobi* zone I am not prepared to share Spath's (1939a, p. 238) opinion on the close relationship between the discussed forms of that zone and *Acanthohoplites aschiltaensis* (Anthula). Even *"Parahoplites" hanovrensis* Collet (1907, p. 225–226, Figures 8 and 9; Pl. 8, fig. 5), which has a suture with symmetrical saddles and a nearly symmetrical first lateral lobe, and in a side view superficially resembles *Acanthohoplites hesper* (Pl. 20, figs. 1–6), clearly shows a depressed venter.

Taking, therefore, *Ammonites milletianus* d'Orbigny (1840, p. 263, Pl. 77, figs. 1–5), a form with a markedly depressed ribbing in the adolescent and adult stages, as the genolectotype of *Hypacanthohoplites*, and assuming that Collet's types with a similar ventral costation belong in this genus, one cannot fail to notice the essential difference between the Acanthohoplitinae of the *nolani* and *jacobi* zones—a marked and rapid development of forms with a more or less depressed venter in the latter zone.

The true parahoplitan and acanthohoplitan faunas seem to terminate with the end of Aptian time, and apparently the natural lower boundary of the Albian is below the strata with numerous forms that show a rapidly progressing ventral depression and ventral interruption of the ribbing.

The older interpretations of the Aptian-Albian zonal sequence and the boundary postulated by Spath, and a more recent interpretation of Muller and Schenck are given in the tables below.

The Aptian-Lower Albian zones were summed up by Spath (1921, p. 311) as in Table 3.

Later, Spath (1923b, p. 146–149), noting that *Ammonites furcatus* Sowerby of pre-*deshayesi* age differs from *Dufrenoya dufrenoyi* (d'Orbigny), replaced the *furcatus?* by the *martini* zone. This interpretation, with the emphasis on the Aptian ammonoid zones of England, is essentially as in Table 4.

Part of the table, from the *deshayesi* to the *mammillatum* zone, prepared by Muller and Schenck in 1943, is given in Table 5.

As has been pointed out above, the Cheloniceratidae are absent from the Cretace-

---

[12] It should be noted, however, that a similar but much lesser depression is observed in a Caucasian species, and in certain varietal forms of a species found in Arizona, of the stratigraphically earlier *Paracanthohoplites* which I have placed in the Acanthohoplitinae.

TABLE 3.—*Aptian-Lower Albian Zones*
Spath, 1921

	Zones	Subzones
LOWER ALBIAN	*tardefurcata*	*regularis* *milletianus*
	nodosocostatum	*jacobi* *nolani*
UPPER APTIAN (Gargasian)	*subnodosocostatum* *furcatus?*	
LOWER APTIAN (Bedoulian)	*deshayesi* *weissi* *bodei*	

TABLE 4.—*Aptian-Lower Albian zones*
Spath, 1923

	Zones	Subzones
LOWER ALBIAN	*tardefurcata*	*regularis* *milletianus* *schrammeni*
	nodosocostatum	*jacobi* *trautscholdi* *abichi*
UPPER APTIAN (Gargasian)	*subnodosocostatum* (Parahoplitan)	*aschiltaensis* *nutfieldensis*
	martini (Tropaeuman)	*tovilense* *bowerbanki* *hillsi*
LOWER APTIAN (Bedoulian)	*deshayesi* (Parahoplitoidan)	*consobrinoides* *hambrovi* *weissi* *bodei*
	(Parancyloceratan)	*bidentatus* *rude* *sparsicosta*

ous sequence of Arizona, and the old world stratigraphic tables with the zones established on the cheloniceran species apply to the current work only by inference. Table 6 presents my interpretation of the Aptian-Albian zonation based essentially on the evidence of the ammonitic succession in southeastern Arizona and on the revision of the family Parahoplitidae discussed in the paleontological part of the paper. This table differs from the tables of Spath and Muller-Schenck in the following: 1. The true parahoplitan or *melchioris* zone is introduced at the base of the Gargasian. 2. The *martini* zone is replaced by the *dufrenoyi* zone. 3. The Clansayan division of the Aptian is recognized as corresponding essentially to the *Acanthohoplites* age. 4. The Aptian-Albian boundary is drawn between the *nolani* and *jacobi* zones. 5. The *trinitensis* zone is placed *below* the *tardefurcata* zone.

TABLE 5.—*Aptian-Lower Albian zones*
Muller and Schenck, 1943

	Zones	Subzones
ALBIAN	*Douvilleiceras mammillatum*	*Douvilleiceras inaequinodum* *Douvilleiceras mammillatum*
	Sonneratia trinitensis	
	Leymeriella tardefurcata	*Leymeriella regularis* *Leymeriella tardefurcata* *Leymeriella schrammeni*
	Acanthohoplites jacobi	*Acanthohoplites jacobi* *Acanthohoplites nolani*
APTIAN	*Cheloniceras subnodosocostatum*	*Acanthohoplites aschiltaensis* *Parahoplites nutfieldensis*
	Cheloniceras martini	*Ammonitoceras tovilense* *Tropaeum bowerbanki* *Tropaeum hillsi*
	Deshayesites deshayesi	*Deshayesites consobrinoides* *Cheloniceras hambrovi* *Deshayesites weissi* *Deshayesites bodei*

CORRELATION WITH THE APTIAN-ALBIAN SEQUENCE IN THE CAUCASUS

The ammonite fauna of the Lowell formation is most readily comparable to the Lower Cretaceous assemblages described by Anthula (1899) and Kazansky (1914) from Daghestan, North Caucasus, and by Sinzow (1908) both from the Mangyshlak Peninsula of the Transcaspian Region and the Caucasus. Later Renngarten (1931) traced the Daghestan fauna further westward along the northern front of the principal range of the Caucasian Mountains from Ingushetia, through Nalchik, to Kislowodsk. He has pointed out: (1) the great development of the strata of Clansayan age throughout the region; (2) the westward change from a deep sea to a littoral facies; (3) the gradual impoverishment of ammonite faunas west of Daghestan.

Unfortunately, the entirely paleontological monographs of Anthula, Sinzow, and Kazansky provide little material on stratigraphy of the described faunas. Anthula and Sinzow based their condensed interpretations on the much older and very general explorations of Abich and others. Kazansky briefly visited Daghestan but published only lithologic characteristics of the strata at various localities. Consequently,

TABLE 6.—*Correlation of the Lower Cretaceous index zones of Arizona with the standard ammonite succession*

	Standard zones	Southeastern Arizona
ALBIAN	*Stoliczkaia dispar*	Patagonia group, Molly Gibson formation. *Stoliczkaia scotti, S. patagonica, S. excentrumbilicata*
	Mortoniceras inflatum	Possible non-sequence
	Euhoplites lautus	Bisbee group Cintura formation
	Hoplites dentatus	Mural limestone. *Orbitolina texana*
	Douvilleiceras mammillatum	*Douvilleiceras? muralense*
	Leymeriella tardefurcata	Lowell formation, Pedregosa member. *Nerinea* sp. Perilla member. *Trigonia mearnsi*
	Sonneratia trinitensis	No ammonites
	Hypacanthohoplites? jacobi	
APTIAN Clansayan	*Immunitoceras nolani*	Quajote member, div. 3b. *Immunitoceras immunitum, Acanthohoplites berkeyi, A. erraticus, A. hesper, A. impetrabilis, A. teres*
	Acanthohoplites aschiltaensis	Quajote member, div. 3d. *Acanthohoplites schucherti, Beudanticeras victoris*
Gargasian	*Dufrenoya dufrenoyi*	Cholla member, div. 5g. *Dufrenoya justinae* Joserita member, div. 8d. *Dufrenoya joserita*
	Parahoplites melchioris	Joserita member, div. 8e. *Paracanthohoplites meridionalis* Pacheta member, div. 9c. *Kazanskyella arizonica, Sinzowiella spathi*
Bedoulian	*Deshayesites deshayesi*	Morita formation

the stratigraphic position of the Daghestan ammonites can be established only by comparing "occurrences" with the lithologic outline. This situation has been ameliorated to a certain extent by Renngarten's reconnaissance in the eastern part of Daghestan. Therefore, the present interpretation of the Aptian-Albian ammonite succession in the Caucasus is based on Renngarten's analysis of the carefully studied sequence in Ingushetia, his observations east and west of the studied area, and on whatever could be gleaned from the perusal of Kazansky's monograph.

The *Deshayesites deshayesi* zone in Daghestan is characterized by *D. dechyi* (Papp), (Table 7), which differs from Leymerie's species by a higher whorl section and ribs which are stronger at the umbilical edge (Renngarten, 1926, p. 32). The *melchioris*, the *dufrenoyi*,[13] the *aschiltaensis*, and the *nolani* zones are all well indicated. In the Albian part of the sequence *Hoplites dentatus* (Sowerby) and *Mortoniceras inflatum* (Sowerby) are present.

The sequence in Ingushetia differs in the absence of the parahoplitan faunas. The *dufrenoyi* zone is represented by the strata with *Cheloniceras martini* var. *caucasicum* (Anthula). The equivalent of the *aschiltaensis* zone yielded no ammonites (unless the stratigraphically lower marls with *Acanthohoplites? subpeltoceroides* Sinzow belong in it), but the *nolani*, the *dentatus*, and the *inflatum* zones are all characterized by the index ammonites.

At Nalchik, Renngarten mentions the succession of *Deshayesites dechyi* (Papp), *Pharahoplites melchioris* Anthula, *Immunitoceras nolani* (Seunes), *Hoplites dentatus* (Sowerby), and *Mortoniceras inflatum* (Sowerby).

At the base of the section at Kislowodsk is a coarse red sandstone with *Oppelia* (*Aconeceras?*) *trautscholdi* Sinzow followed by dark-gray arenaceous marls with *Deshayesites dechyi* (Papp) and yellow sandstones with *Cheloniceras cornuelianum* (d'Orbigny). Gray marls above, about 600 feet thick, contain several species of *Parahoplites, Cheloniceras*, and *Acanthohoplites*. The overlying brown sandstone is characterized by *Immunitoceras nolani* (Seunes). In the Albian part of the sequence the fauna is poorly represented except in the topmost black clay, from which *Mortoniceras inflatum* (Sowerby) was collected.

Table 7 shows a correlation of the Caucasian sequence with the standard zones used for southeastern Arizona.

CORRELATION WITH THE SEQUENCE IN THE QUITMAN MOUNTAINS, TEXAS

The sequence of the Cretaceous strata in the Quitman Mountains of Texas has been discussed and illustrated by Gayle Scott (1940, p. 978, Pl. 55, and p. 980). In those stratigraphic units characterized by ammonite faunas the equivalent strata of the Travis Peak formation with *Dufrenoya* are followed by thin-bedded limestones that contain a *Sonneratia* fauna, then by blue marls in which occur several species of *Douvilleiceras* and which grade upward into gray marls with *Douvilleiceras mammillatum*, and finally by the great series of limestones with *Orbitolina texana* and the rudistids of the Glen Rose formation.

The *Sonneratia* fauna of Texas appears to be unrepresented in Arizona. However, its approximate stratigraphic position in the Lowell formation may be established on the following ground. In the fall of 1940, while geologizing in the Quitman Mountains under the kind and instructive guidance of Dr. Scott, I observed in Mayfield Canyon well-preserved specimens of *Trigonia mearnsi* in the blue marls with *Trinitoceras rex* Scott (1940, p. 1017, Pl. 60, figs. 1–3), which stratigraphically are much higher than the beds with *Sonneratia*. In the Lowell formation of Arizona this *Trigonia* has a well-defined vertical range. It occurs in division 2b of the Perilla

---

[13] Kazansky (1914, p. 106) identified *Dufrenoya "furcata"* of Daghestan with d'Orbigny's species.

TABLE 7.—*Correlation of the Lower Cretaceous index zones of the Caucasus with the standard ammonite succession*

Standard zones	Daghestan Interpreted from Kazansky, 1914, and Renngarten, 1931	Ingushetia Modified from Renngarten, 1931
Stoliczkaia dispar	Not represented	Not represented
Mortoniceras inflatum	Black arenaceous shale. *Mortoniceras inflatum*	1. Light-gray marl and dark-gray shale, 10 feet. *Mortoniceras inflatum*
Euhoplites lautus		
Hoplites dentatus	Light-gray marl. *Hoplites dentatus*	2. Dark-gray shale, 43 feet. *Hoplites dentatus*
Douvilleiceras mammillatum		3. Unfossiliferous dark clay, 210 feet.
Leymeriella tardefurcata		
Sonneratia trinitensis		4. Dark-gray clay with argillaceous sandstone, 525 feet. *Thetironia caucasica*
Hypacanthoplites? jacobi		
Immunitoceras nolani	Arenaceous marl. *Immunitoceras nolani, Acanthohoplites aschiltaensis, Ac. abichi, Ac. bergeroni*	5. Dark-gray clay, 33 feet. *Immunitoceras nolani*
Acanthohoplites aschiltaensis		6. Gray marly glauconitic sandstone, 56 feet. *Nuculana scapha*
Dufrenoya dufrenoyi	Dark Clay. *Dufrenoya dufrenoyi*	7. Gray marl, 85 feet. *Acanthohoplites? subpeltoceroides* 8. Dark arenaceous clay, 50 feet. *Cheloniceras martini* var. *caucasicum*
Parahoplites melchioris	Marly sandstone with concretions. *Kazanskyella daghestanica, Parahoplites melchioris*	?
Deshayesites deshayesi	Gray arenaceous marl. *Deshayesites dechyi*	9. Greenish-gray glauconitic marly sandstone, 155 feet. *Cheloniceras seminodosum, C. cornuelianum* 10. Yellow-gray marly sandstone, 7 feet. *Deshayesites dechyi, Oppelia (Aconeceras?) trautscholdi*

member about 162 feet above the *Acanthohoplites* zone which terminates in division 3b of the Quajote member. In the Arizona sequence, therefore, the equivalent position of the *Sonneratia* fauna of Texas should be between the *Acanthohoplites* and the *Trigonia mearnsi* zones.

In a correlation table of Texas, Arkansas, Louisiana, and New Mexico, Scott (1940, p. 980) showed a non-sequence between the *Dufrenoya justinae* and the *Sonneratia trinitensis* zones. The gap, according to the table, corresponds to the *Leymeriella regularis* zone.[14] I think, however, that the *Leymeriella* zone is *above* and not *below* the beds with *Sonneratia* of the Quitman Mountains. Scott (1940, p. 974) himself asserted that: "In the thick section of the southern Quitman Mountains *Sonneratia* occurs several hundred feet below the zone of *Douvilleiceras mammillatum*." He refers to Spath's statement[15] that "In Mangyshlak Peninsula *Sonneratia* occurs well above the zone of *Leymeriella*, but *Douvilleiceras* is not found in that area." It should be mentioned, first, that several forms from Mangyshlak described by Sinzow under *Sonneratia* come from different, in part unestablished, localities and horizons, and second, that they have little in common with the species from the Quitman Mountains.

Assuming the correctness of the above interpretations, it is to be pointed out that Ninety One Hills area has 182 feet of strata between the *Trigonia mearnsi* zone and the thinner-bedded basal strata with *Orbitolina texana* of the Mural limestone. Scott believes (personal communication, June 19, 1943) that in the Quitman Mountains *Trinitoceras rex* occurs a little below *Douvilleiceras spathi* and considerably below the *Douvilleiceras mammillatum* zone. Thus far only a doubtful specimen which may belong in *Douvilleiceras* has been collected from the basal part of the Mural limestone (Pl. 26, figs. 1–2). However, more than 400 feet of strata between the beds with *Acanthohoplites* and those with *Orbitolina texana* provides ample space for the stratigraphic equivalents of the *jacobi, trinitensis, tardefurcata,* and *mammillatum* zones.

CORRELATION WITH THE STRATIGRAPHIC UNITS OF TEXAS

Few index species are common to the Lower Cretaceous of Texas and the Lowell formation of Arizona: (1) *Dufrenoya justinae* of the Cholla member, which occurs in Texas in the Travis Peak formation and in the lower part of the Cuchillo formation. Within the Lowell formation *Dufrenoya* appears to range from the Joserita member, division 8d, to the Cholla member, division 5g. (2) *Trigonia mearnsi* of the Perilla member is well represented in Texas in the upper part of the Cuchillo formation in the Quitman Mountains. (3) *Trigonia stolleyi* of the same member is found in the Glen Rose formation of Texas.

Of a particular interest is the presence in the Cretaceous strata of Texas of certain well-preserved Trigoniae which have never been described. Three specimens of these Trigoniae, labeled as "*Trigonia* sp. 1, Travis form., Loc. 1804, Panther Creek, on road from Travis Peak to Burnet, between 18th and 19th mile posts from Burnet,"

---

[14] This interpretation was also reproduced by Muller and Schenck (1943, p. 272–273, fig. 6. *Compare* Table 5).

[15] Based on A. Nacki's paper: "Zone mit *Hoplites* (*Leymeriella*) *tardefurcatus* Leym., an Mangyschlak," Annuaire Géologique et Minéralogoque de la Russie, vol. 14, 1912. This publication is not available in the United States. In Nacki's section, *Sonneratia* sp. occurs in the *tardefurcata* zone, and *S.* cf. *dutempleana* above it (*fide* Kilian, 1915, p. 171).

are in the U. S. National Museum. The species represented by one of these examples is very close to *Trigonia reesidei* of this paper and all three specimens belong in the *abrupta* group. I was unable to place these Trigoniae stratigraphically, either through the literature or in discussions with geologists. However, in the same collection and similarly labeled are Trigoniae of the *excentrica* group. In the Lowell formation, Trigoniae of the latter group range from the Saavedra member to the Quajote member, which fact by itself is not of much assistance, but Hill (1893, p. 30, Pl. 4, fig. 3) described (as *Pholadomya lerchi*) a poorly preserved Trigonia of the *excentrica* group from the conglomerate at the base of the Comanchean series on Sycamore Creek, near the crossing of the Burnet and Travis Peak roads. I do not know the exact relation of these two localities which seem to be close topographically. What I want to point out is this. With the possible exception of the Trigoniae of the *abrupta* group, no species of ammonites and lamellibranchs from the two lower members of the Lowell formation—the Pacheta and the Joserita (of 220 feet aggregate thickness)—is known in Texas. The upper limit of *Trigonia reesidei*, to which one of the Trigoniae of the *abrupta* group bears a strong resemblance, is in the Arkill limestone, Saavedra member, division 71, whereas in the conglomeratic part of the stratigraphically lower Espinal grit the same species occurs with numerous specimens of Trigoniae of the *excentrica* group. Whether the related *abruptae* in Texas were redeposited in the same basal Comanchean conglomerate as the one mentioned by Hill and which, according to Adkins (1932, p. 274, 284), was spread on the Wichita Paleoplain at different ages, or whether they came from a stratum above it is not known. Judging by the preservation the specimens were not collected from a conglomerate. However, they should not be older than the Arkill limestone of Arizona because of the total absence of the other older faunas of the Pacheta and Joserita members in Texas. If these postulates are correct, the two lower members of the Lowell formation are older than any paleontologically characterized Aptian stratigraphic unit of Texas.

The strata from the Saavedra member to the Perilla member should correspond in age to the Travis Peak formation and to the lower division of the Cuchillo formation of the Quitman Mountains. The Perilla member stratigraphically corresponds approximately with the middle part in the upper division of the latter formation. The transition to the equivalent of the Glen Rose formation is in that part of the Arizona sequence where the yellow and brown beds of the Pedregosa member begin to alternate with the basal gray strata of the Mural limestone.

The Mural limestone is composed generally of thinner-bedded gray limestone with *Orbitolina texana* both below and above the massive limestone with the rudistid reefs. Its age, therefore, is determined by the vertical range of *Orbitolina texana*. The Mural limestone is stratigraphically equivalent to the Glen Rose formation.

Strata of Glen Rose age, lithologically different from those of the Mural limestone, are present also on the west side of the Huachuca Mountains. I have collected typical specimens of *Ostrea ragsdalei* Hill in the Park Canyon above the highway where the canyon cuts into the mountains. The upper part of the section, below the highway, has not been examined.

In the Patagonia Mountains, the Cretaceous sequence—the Patagonia group—is over 5000 feet thick. The only fossiliferous unit located, the arenaceous limestone beds with several species of *Stoliczkaia*, is in the Molly Gibson formation in the upper part of the sequence. Quite possibly the entire succession from the Walnut to the Grayson-Del Rio of Texas is contained within the Patagonia group. Apparently this is a marginal facies which overlaps the Paleozoic strata to the north, and in which the equivalents of the Cintura formation of the Bisbee area and of the Buda limestone of Texas may also be represented. However, there may have been interruptions in deposition within the lower 2000 feet of the group below the only known fossiliferous formation.

THE MALONE CONTROVERSY

Since the Malone controversy bears directly on the research in the Cretaceous stratigraphic paleontology of Arizona, its history is briefly summarized.

The invertebrate fauna from the Malone Mountains and the Malone Hills between Finlay and Sierra Blanca in Hudspeth County, Texas, was described and illustrated by Cragin in 1905. As in the Cretaceous faunas of southeastern Arizona, the Malone index fossils are pre-eminently ammonites and Trigoniae. Cragin interpreted the Malone fauna as Jurassic, chiefly on the basis of ammonites, some of which he regarded as conspecific with the forms previously described by Castillo and Aguilera (1895) from Catorce, Mexico. He regarded the Malone Trigoniae, as a whole, as related more to the Jurassic than to the Cretaceous species. However, in his opinion the presence of *Ptychomya*—a pre-eminently Cretaceous lamellibranch genus—in the same assemblage offset the strongly Jurassic aspect of the fauna and probably made it of a later age than indicated by the ammonites (Cragin, 1905, p. 19).

Subsequently Uhlig (1910, p. 69) upheld the Jurassic age of the ammonites described by Cragin. He did not fail to notice, however, the strikingly Cretaceous features of some of the Malone Trigoniae, and to him belongs the concept that Trigoniae of that type might have persisted from Late Jurassic into Early Cretaceous time.

Twenty one years after the publication of Cragin's paper Kitchin (1926, p. 454-469), the outstanding student of the fossil Trigoniae, analyzed in a brilliantly presented discussion the Malone lamellibranchs and especially the Trigoniae as described and illustrated by Cragin. Admitting the Jurassic age of the Malone ammonites on the authority of Uhlig and a later check-up by Spath (Middle Kimmeridgian—probably Upper Kimmeridgian—possibly Tithonian), Kitchin could not assign the same age to the Malone Trigoniae which, he maintained, had a definitely Cretaceous aspect and relationship. He also pointed out that these Trigoniae are too specialized for the geological longevity presumed by Uhlig. Accepting an earlier suggestion of Burckhardt (1912, p. 218, footnote) as to the possibility of faunas of two different ages in the Malone strata (ammonites—Jurassic, and Trigoniae—Cretaceous), Kitchin postulated the presence of both Jurassic and Cretaceous deposits with an undiscovered non-sequence within the known collecting grounds in the Malone area.

Analyzing the all important *Trigonia vyschetzkii* Cragin, Kitchin pointed out that the ontogenic development of the escutcheonal and areal ornamentation in some of Cragin's specimens referred to this species indicated Early Cretaceous age. Although Kitchin postulated that none of the Malone Trigoniae is quite identical with described species of the South Andean realm, Weaver's (1931, p. 110–115) more recent studies have demonstrated that in Argentina the species of the *Trigonia transitoria* group, which morphologically closely resemble the pseudo-quadrate Trigoniae of Malone, are all found in the Lower Cretaceous strata with Valanginian and Hauterivian ammonites. No less impressive is Kitchin's emphasis on the importance of a correct discrimination in the early individual development of the homoeomorphic species of Trigoniae belonging in the *v-scripta–vau* groups of India, South Africa, South America, Mexico, and Texas for an evaluation of their interrelation and age. The presence of such genera as *Ptychomya* and *Eriphyla* intensifies his age interpretation of the Malone lamellibranchs. According to Weaver (1931, p. 116–127), in Argentina these genera occur only in Cretaceous strata. Probably less convincing is Kitchin's suggestion of the degeneracy of the Malone Trigoniae affiliated with the Costatae.

Kitchin's conclusions of his analysis are as follows: What Cragin regarded as the Malone Jurassic formation is in reality a series of separate units. A few Jurassic horizons are present, but not necessarily as a continuous sequence. Above the Jurassic strata, and with as yet undetermined relationship to the latter, are Lower Cretaceous beds, probably of Valanginian age, with Trigoniae affiliated with those in the Cretaceous strata of the southern realm.

Adkins (1932, p. 254–256, and 286–288) accepted Kitchin's "inferred unconformity" between the Jurassic strata with ammonites (the Malone formation proper) and the Cretaceous strata with Trigoniae (the Torcer formation) at Malone.

Albritton (1938, p. 1747–1806) has shown that the Malone Jurassic sequence is separable into a lower division, essentially made up of clastic rocks (arenaceous shale, siltstone, and occasionally conglomerate and limestone), to which all Trigoniae and the majority of ammonites described by Cragin are restricted, and an upper, more calcareous division with *Kossmatia*. Detailed zonation was impossible because of the scarcity of ammonites. Of the described sections, section No. 6 shows the presence of *Trigonia vyschetzkii* and *Idoceras clarki* in the same stratum. In unit 27 of section No. 7 the same ammonite species occurs with *Trigonia calderoni* of Cragin; this is by the way, the only stratum from which this species of Trigonia was reported. In section No. 8 *Tr. vyschetzkii* is above the beds with ammonites, but *T. proscabra* occurs both below and above. In section No. 12 the beds with *T. vyschetzkii* and with *Idoceras clarki* are separated by a coarse conglomerate. In section No. 23 *Ptychomya stantoni* and *T. vyschetzkii* are above the beds with ammonites.

Cragin's ammonites species are all autochthonous, not identified yet in other localities,[16] although their genetic relation to the better known species, and especially to those of Mexico, cannot be doubed, as has already been pointed out by Burckhardt (1912, p. 218). Albritton's direct correlation of the lower division of the Malone

---

[16] According to Albritton (1937, p. 391–412) Cragin's identification of some of the Malone ammonites with the species described by Castillo and Aguilera (1895) from Catorce, Mexico, is incorrect.

formation with the *Idoceras-Haploceras* shales of Mexico is necessarily based on the common genera. Burckhardt's three Mexican species described by Albritton from Malone—*Nebrodites nodosocostatus, Aspidoceras laevigatum,* and *Kossmatia zacatecana*— have not been illustrated except for very schematically drawn whorl sections, and therefore their stratigraphic significance cannot be evaluated (Albritton, 1937, p. 402, 407, 410, Pls. 7–9. *Compare* Spath, 1939b, p. 207).

The term "Torcer formation" as employed by Albritton for the Cretaceous part of the Malone sequence has an entirely different meaning from that originally assigned to it by Adkins. The Torcer formation of Adkins was a provisional unit for the strata with Cragin's Trigoniae, supposedly *above* the hypothetical disconformity postulated by Kitchin that was to separate it from the older beds with Jurassic ammonites. To Albritton the Torcer formation represents 400 feet of strata made up of clastics and limestones without Trigoniae but with Cretaceous foraminifera and with a fragmental ammonite in the basal division of sandstones and conglomerates which was identified as *Neocomites* cf. *indicus* Uhlig (Albritton, 1938, p. 1764–1766; 1937, p. 408), all well above the strata with both the Trigoniae and the ammonites described by Cragin.

Thus, the fauna of the Malone beds is distinguished from all other known late Jurassic faunas of admittedly same age in the presence of much discussed Trigoniae, the Cretaceous aspect of which has been recognized by Uhlig, Burckhardt, and Kitchin, and is substantiated by the present research. *Trigonia vyschetzkii* Cragin, as defined in this paper, strongly resembles not only certain forms of the *T. transitoria* group of South America, all species and varieties of which are restricted to the Valanginian and Hauterivian, as has been shown by Weaver (1931, p. 56, and 244) in the Cretaceous strata of Argentina, but also certain Pseudo-Quadratae of the Arizona Aptian. Yet, there is a much closer similarity between the Trigoniae of the *v-scripta—vau* groups of Malone and of southeastern Arizona. If the survival of such highly specialized types of Trigoniae as the Pseudo-Quadratae from Late Jurassic into Early Cretaceous seemed quite problematical, it is more difficult to account for the presence of nearly identical forms in the Aptian strata. Albritton's (1938, p. 1769) designation—"pseudo-Cretaceous pelecypods of the Malone fauna"—seems not very appropriate, at least as far as Trigoniae are concerned, since the Trigoniae are of characteristically Cretaceous types which, barring their presence in an exceptional association with Late Jurassic ammonites of Malone, are restricted elsewhere to Cretaceous strata. It is the time of origin and early development of such Trigoniae that is involved in the problem. In none of the sections studied by Albritton were the Kimmeridgian ammonites found above the beds with Trigoniae. *Kossmatia,* located in the upper division of the Malone formation above these beds, is a Tithonian-Berriasian genus (Roman, 1938, p. 326). The fragmentary *K. aguilerai* (Cragin, 1905, p. 105, Pl. 28, figs. 1–2) was considered Tithonian by Uhlig and Late Tithonian (Portlandian supérieur) by Burckhardt (1930, p. 83). The specimen identified by Albritton (1937, Pl. 7, fig. 3) as the Mexican *K. zacatecana* Burckhardt has been illustrated only by crude drawing of a whorl section. The generally poor, fragmental preservation of the ammonites and Trigoniae in the Malone area is a result of bad weathering. Probably the method systematically employed in the

Ninety One Hills area of southeastern Arizona—*i.e.*, digging deep trenches through apparently unfossiliferous strata—, which was so fruitful in obtaining good specimens of ammonites not indicated at the surface, would add to the knowledge of the upper part of the Malone formation.

MALONE FACIES

Besides the similarity in the discussed Trigoniae there is also a general facial resemblance between the faunal asemblages observed in the Malone strata and in the middle part of the Lowell formation. Closely associated with the Trigoniae in both units are serpulids, vermetids, corals of "*Astrocoenia*" type, and especially the lamellibranch element composed of numerous lucinids, astartids, small ornamented Gervilliae, and forms usually referred to *Pleuromya*. The conformity of species and the parallelism of the entire assemblages are considerable. In the Cholla member of the Lowell formation this type of fauna is coexistent with the Trigoniae of the analyzed groups. This facies is also represented in Mexico where a similar fauna with the Trigoniae identified as *T. vyschetzkii* and *T. calderoni* occurs in the Lower Cretaceous strata of the Laguna District, Coahuila (Kellum, 1936b, p. 1055–1056, 1066–1067), and, without the pseudo-quadrate Trigoniae, at Catorce, San Luis Potosi (Castillo and Aguilera, 1895). The presence of pseudo-quadrate Trigoniae in northern Sonora is of special interest (Burckhardt, 1930, p. 150).

This characteristic facies has not been reported south of Catorce. In Argentina, with all abundance of the pseudo-quadrate Trigoniae in the Lower Cretaceous formations, the forms which may be connected with the *v-scripta* and *vau* groups are very remote from the species that occur in Texas, Mexico, and Arizona (Burckhardt, 1903, p. 72, Pl. 13, fig. 1; Weaver, 1931, p. 261; Pl. 25, figs. 131–136; Pl. 27, fig. 150). On the other hand, in Peru, as in Catorce, the materials with Trigoniae referable to these groups (Lisson, 1930) seem to contain no pseudo-quadrate Trigoniae.

Kitchin quoted Uhlig on the similarity of Trigoniae in the isolated areas of the so-called southern trigonia sea to the effect that: "One can have no doubt not only that this fauna is distributed through the whole South Andean region from Patagonia to Malone, but that it is essentially identical with the Uitenhage *Trigonia* fauna of South Africa," and commented: "We must remember, however, that he [Uhlig] was content to regard the fauna as Neocomian in South Africa, Infravalanginian in Patagonia, and Jurassic (perhaps so old as Kimmeridgian) in Texas." When introducing the theoretical presence of both the Jurassic and the Cretaceous strata in the Malone succession Kitchin added: "This southern bivalve fauna, so widely distributed, occurs in strata deposited certainly within the period from the beginning of Cretaceous time to late Valanginian; it is probably entirely Valanginian" (Kitchin, 1926, p. 466–467).

With Trigoniae of the "southern type" found in the upper division of the Aptian in southeastern Arizona, and the Jurassic age of the Malone Trigoniae corroborated, the migration "through ages" postulated by Uhlig would require a time from Kimmeridgian to Gargasian.

Kitchin's error perhaps was not in the evaluation of the ontogenic and phylogenic development of the southern Cretaceous Trigoniae, which in the main is well worked out, considering the available paleontological material and geological information, and is even now a solid foundation for any further research in this line but in the misinterpretation of stable and variable characters in their relation to the scale of time. The features that were supposed to indicate a rapid and geologically short specialization, as for instance the resolution of the escutcheonal costellae in the Pseudo-Quadratae, appear to have been introduced earlier and remained, with slight variations, through the Lower Cretaceous.

Continuous intercommunication from the southern part of North America via South America to Africa, and supposedly even to India, through all the time involved, seems improbable. A more plausible postulate is that the essential morphological characters of the pseudo-quadrate, *v-scripta,* and *vau* groups were already established at Malone time. Kitchin probably was right to restrict the actual communication between South America and Africa to the Neocomian. With the intermigration thus restricted, one must admit the existence of an asylum north of the equator in the western hemisphere in which the Malone facies persisted, and the slow development of the homoeomorphic Trigoniae continued, from Late Jurassic to late Aptian time.

RELATION BETWEEN THE MALONE AND THE LOWER CRETACEOUS TRIGONIAE

In the fall of 1940, after studying Cragin's collections of fossils from Malone, which are preserved in the U. S. National Museum at Washington, I made a brief visit to the Malone and the Quitman Mountains areas in Texas under instructive leadership of Dr. Gayle Scott and in company of Dr. J. B. Reeside, Jr. and Dr. R. W. Imlay.

At Malone the topotypes for all types of Cragin's *"Trigonia vyschetzkii"* were collected in the same stratum with his *"Trigonia calderoni."* No other Trigoniae were found at the same horizon. Beds closely associated but slightly lower, contained poorly preserved ammonites; a few specimens related to *Idoceras schucherti* were identifiable.

The reader is refered to the paleontological part of the paper for a critical description and comparative analysis of the Malone Trigoniae which are morphologically related to the species from the Cretaceous strata of southeastern Arizona. Here are presented a few observations of general character relevant to the subject.

It is necessary to state at the outset that Cragin's types described as *"Trigonia vyschetzkii"* do not constitute a conspecific assemblage. Only two of his four illustrated specimens can unreservedly be placed in the Pseudo-Quadratae. It is evident from Cragin's description that his diagnosis of the species was made essentially from one of these two closely related examples, which I have selected as the lectotype of *Trigonia vyschetzkii* Cragin (1905, Pl. 8, fig. 2, only). The two other specimens mentioned above (Cragin, 1905, Pl. 9, figs. 1 and 2) seem to have certain characteristics that considerably deviate from the lectotype. These two forms are discussed under a different name.

Another important species that belongs in the pseudo-quadrate Trigoniae of Malone, although not figured by Cragin, is illustrated in this paper and referred to as *Trigonia dumblei,* nom. nov. (Pl. 12, fig. 4). Originally the type was mistaken by

Cragin for *T. taffi* Cragin, and its type locality was, also erroneously, given as Bluff Mesa. Later Cragin corrected the reference of the finding place, but incorrectly identified the type with *T. vyschetzkii*.

If such characteristics as the elongate and posteriorly attenuated shell, the well-developed encroachment of the areal costellae upon the posterior part of the flank, the presence of an inframarginal band, and the wider outer area actually indicate, as Steinmann thought, a closer relationship with the Clavellatae,—the conclusion that the Cretaceous Pseudo-Quadratae of Argentina are of a much older aspect than those of Malone is inescapable. Steinmann's view, however, was not shared by Kitchin, and indeed, according to Weaver (1931, p. 56), the elongate forms with clavellate characteristics, like the Argentina specimens referred to *Trigonia transitoria* and its variety *vacaensis*, are found to be stratigraphically higher (Hauterivian) than the varieties *quintucoensis* and *curacoensis*, and *Trigonia neuquensis* Burckhardt (Valanginian) which have shorter shells and less strongly expressed "clavellate" features. *T. transitoria* Steinmann from the vicinity of the volcano Antuco, Chile, *T. vyschetzkii* of Malone, and the pseudo-quadrate Trigoniae of Arizona are also shorter and less like the Clavellatae.

It is obvious that the degree of the much discussed encroachment of the areal costellae upon the flank is proportional to the length of the shell and is strongly developed only in the species and varieties with longer shells.

The studies suggest that it is the modification and degree of the following morphological characters that are significant in a diagnostic evaluation of the Pseudo-Quadratae: (1) resolution of the areal costellae, (2) pairing of the areal costellae, (3) relative development and tuberculation of the median carina, (4) orientation of the tubercles on the escutcheon.

Species with unresolved areal costellae are *T. transitoria* Steinmann (1882, Pl. 8, fig. 3, only), *T. vyschetzkii* Cragin (1905, Pl. 8, fig. 2), and *T. guildi* (Pl. 12, figs. 1–2). The shell of the first species is the longest of the three. Its outer area is notably wider than the inner area, the untuberculated median carina is nearly altogether superseded by the supramedian furrow, the inframarginal band is present, and the infringement of the areal costellae upon the posterior part of the flank is well developed. In the shorter and posteriorly less attenuated shell of *T. vyschetzkii* the difference in the width of the two areas is less, the median carina is present and has a tendency toward tuberculation in its anterior part, there is no inframarginal band, and the encroachment of the areal costellae upon the flank is insignificant. *T. guildi* differs from the Malone species in being shorter, in having subequal areas, in a strongly developed median carina which is tuberculated for two anterior thirds of its length, and in paired areal costellae—a feature also shared by one of the two syntypes of Steinmann's *T. transitoria* (Steinmann, 1882, Pl. 8, fig. 1).

The resolution of the areal costellae has not been observed in the South American species of the Pseudo-Quadratae. In *T. dumblei* of Malone (Pl. 12, fig. 4) and in *T. resoluta* of Arizona (Pl. 12, fig. 3) the costellae of the outer area (only) are resolved in its anterior part and are paired posteriorly.

Steinmann (1881, p. 261) regarded the oblique arrangement of escutcheonal tubercles as one of the principal features that connect the Pseudo-Quadratae with the

Quadratae. This orientation, however, is to be accepted as already established in the geologically earlier Cretaceous pseudo-quadrate Trigoniae since it occurs in *T. transitoria* Steinmann (1882, Pl. 8, fig. 2), on which the group was based, and in the majority of varietal forms illustrated by Weaver. Only in some varietal examples from the Aptian strata of Argentina is there a tendency toward a rather indefinite transverse arrangement in the posterior part of the escutcheon (Weaver, 1931, Pl. 23, figs. 119, 123). However, in the Jurassic *T. vyschetzkii* these tubercles are set in transverse rows already in the early part of the escutcheon, and a transverse arrangement is again repeated in the posterior part of the escutcheon of the Aptian *T. guildi*, whereas the oblique orientation is observed in the geologically younger *T. mearnsi* (Pl. 13, fig. 2). With the present limited knowledge of the phylogeny and ontogeny of Trigoniae it is debatable whether all this signifies two different lines of descent or a recurrence of characters with a gradual and repeated replacement of the transverse by the oblique escutcheonal ornamentation through the Late Jurassic and Early Cretaceous.

The younger age of the Arizona pseudo-quadrate Trigoniae, as an assemblage, is indicated by the presence of species, rather low in the Lowell formation, in which the resolution of the areal costellae takes place in both areas—a characteristically "quadrate" feature—as in *T. saavedra* (Pl. 13, fig. 5), and also by the general trend of development toward the forms with a habitus more approaching that of the Quadratae (*T. mearnsi*).

More intricate data, however, are involved in discriminating between the Malone and Arizona Trigoniae of the *v-scripta* and *vau* groups in which knowlege of the early individual development of the shell is indispensable for the correct interpretation of a species. The only Malone representative connected with these groups was identified by Cragin (1905, p. 59, Pl. 9, figs. 4–6) as *Trigonia calderoni* (Castillo and Aguilera, 1895, p. 9, Pl. 5, figs. 17–18) from Catorce, Mexico. Since the apical part of the shell, the escutcheon, and the areas are completely destroyed in the fragmentary specimens of Malone which Cragin interpreted as plesiotypes of the Mexican species, and also since Castillo and Aguilera neither described nor illustrated the corresponding parts of their types, it is impossible to say whether the Catorce and the Malone forms are conspecific, nor is it possible to construe their true relation to other species of the *v-scripta* and *vau* groups. It is apparent, however, that in the syntypes of Castillo and Aguilera the spacing of ribs and the relation of their vertical and horizontal sets are more as in the Aptian *Trigonia cragini* (Pl. 13, figs. 6–9) of Arizona than as in Cragin's examples, in which the ribs are spaced appreciably wider.

Fortunately, a fairly complete undescribed topotype is preserved in the U. S. National Museum; it was collected by Stanton at the west base of Truncate Mound in the Malone Hills, "30 feet beneath base of conglomerate" as marked on the label. According to the map prepared by Albritton (1938, Pl. 1) this specimen must have been collected from the lower division of the Malone formation and near his examined sections Nos. 23 and 24, from which *T. calderoni* has not been reported. The only conglomerate indicated in these sections is at the top of the lower division. Thirty feet below the conglomerate would place the horizon from which Stanton's specimen was obtained higher in the sequence than any of the fossiliferous beds recorded by Albritton for the lower division, about 22 feet above the topmost strata with Trigoniae in section No. 23, and 75 feet above the bed with *T. vyschetzkii* in section No. 24. A cast of the specimen which Stanton identified with *T. calderoni* (Castillo and Aguilera), and therefore indirectly with Cragin's forms, is figured in Figures 1 and 2, Plate 14. The distribution of ribs in this topotype is as in the specimens illustrated by Cragin and therefore different from the ribbing of the species from Catorce and southeastern Arizona. But the early development of its intracarinal characters, especially the progressive interrelation of the areas and the ornamentation, is surprisingly similar to that in *T. cragini*. The observed difference is of a minor nature, and probably would not be regarded as more than varietal if these forms were found in close stratigraphic association. However, in the Lowell formation of Arizona, the strata with *T. cragini* (Cholla member, divisions 6f–6a) are immediately succeeded by the beds with *T. kitchini* (Pl. 14, figs. 4–10), which has a more advanced and elaborate escutcheonal ornamentation than has ever been observed in the described Trigoniae affiliated with the *v-scripta* and *vau* groups.

Assuming the correctness of the accepted succession of strata with which the discussed assemblages of Trigoniae are associated: Malone—Neuquen—Bisbee, it is safe to conclude that in the western hemisphere the Trigoniae of the pseudo-quadrate and the *v-scripta—vau* groups are essentially of Cretaceous age. There is, as far as I am aware at present, nothing to contradict a Cretaceous age for the related Trigoniae of Africa. As has been pointed out in the preceding discussion, the presence of Trigoniae with such strong Cretaceous affinities in the strata of Malone age is quite unique. It does not imply a later age than Jurassic for the strata in which these Trigoniae occur in a close association with the Malone ammonites, but it certainly foreshortens the perspective of the time span which separates these strata from the Cretaceous and through which the development of the Trigoniae of the established types continued uninterruptedly. Nothing justifies the condemnation of Trigoniae of the discussed groups as index fossils. However, careful studies are indispensable for a correct evaluation of the constancy, variability, and recurrence of the principal characteristics, which are not always strictly contemporaneous in the different lines of descent of these interesting lamellibranchs, as has already been pointed out by Kitchin (1908, p. 109) in connection with the development of the Pseudo-Quadratae. As Spath (1935, p. 185) remarked in criticizing Kitchin for ascribing a Cretaceous age to five species of Trigoniae which later proved to belong in the Jurassic part of the Kachh (Cutch) sequence:

"... It is time to remind the general palaeontologist again of what workers on ammonites have realized long ago. That is, the unsatisfactoriness of arguments based on the principle that the stage of evolution attained by certain fossils can be used for exact (as opposite to merely approximate) dating of the beds in which they occur."

REMARKS ON AMMONITES

It may seem platitudinous to speak of the advantage achieved in the paleontology of the Late Paleozoic and Early Mesozoic ammonoids through research in the ontogenic development of the shell, suture, and ornamentation which, inaugurated by Hyatt, Branco, and Karpinsky, and followed by the works of their successors, has placed that division of ammonitology on a more or less secure foundation.

The comparative inadequacy of such studies for many of the Late Mesozoic groups of ammonites undoubtedly resulted from the greater diversification of easily perceptible morphological characters in the adult shells, which was readily accepted as sufficient for classificatory needs inasmuch as it seemed satisfactory for the purpose of general stratigraphic correlations. Considerable work on individual development has been done for certain groups of the geologically later ammonites, but very little has been accomplished for others.

Within the limits of research covered in this paper, the lack of such knowledge is especially felt in connection with the studies of the family Parahoplitidae. It is not an exaggeration to state that the diagnoses for the greater part of the described parahoplitan species have been based either on the last, or on the last available whorl, without sufficient acquaintance with the earlier development of the shell. To wit: in the late stage of growth of many representatives of the Parahoplitidae, the specific and sometimes even generic characteristics are often obliterated by convergence to such an extent as to render the diagnoses established exclusively on the mature whorls practically valueless for the purpose of study. Another example: a reference to the costation in the Acanthohoplitinae as provided with ventral, lateral, or umbilical tubercles is meaningless unless the stage of development is indicated, because in the acanthohoplitan ornamentation the tubercles, or the thickenings on the ribs, gradually migrate with the growth of the shell from the peripheral margin of the whorl to the umbilical edge, and the degree of this process, its acceleration or retardation, varies in different species. The development of the suture also is of considerable assistance in discriminating among the closely affiliated genera of the Parahoplitinae, or for safe guidance in the identification of those acanthohoplitan species which tend, at certain stages of growth, toward convergence with parahoplitan or cheloniceran forms.

It was impossible to make an orderly analysis of the Parahoplitidae of southeastern Arizona without certain revisionary work on this family, designation of holotypes, and selection of genotypes. This part of the research was handicapped by the unavailability of the described European types. For this reason I preferred to base some of the introduced genera on the studied American material rather than to speculate on the insufficiently characterized and illustrated congeneric European forms.

My studies of these ammonites depended on material collected from strata of rapidly changing lithologic nature in which the preservation of fossils also varies considerably. The examples discussed under the Deshayesitinae probably are the least satisfactorily represented both in quality of preservation and in numbers, and therefore least satisfactorily analyzed. However, with the imperative restriction of the limits of the Parahoplitinae, the accepted association was the only one that seemed plausible.

It will be noted that in certain cases, whenever more or less abundant and better-preserved material was available, I preferred to base the introduced species on syntypes rather than on a holotype. Designation of a single specimen to represent a species may be quite satisfactory in the paleontology of gastropods (*compare* J. Brookes Knight, 1941, p. 10). To know an ammonite adequately it is necessary to dissect the specimen for the study of its inner whorls. The chance of finding sets of individuals that died in childhood or at different stages of growth, as J. P. Smith (1932, p. 20) suggested, is very slight, and, as he admitted, even then, with a few possible exceptions, there can be no assurance of identity with the mature specimens that have never been analyzed. However, a thoroughly dissected ammonite is no longer a good museum specimen. By breaking a few of the available examples that appear identical, the interrelation of younger and older whorls is established, and in this way the undissected specimens may be added with a fair degree of certainty to the group of syntypes for a more complete illustration of the species. The disadvantage is in dealing with more than one specimen in the subsequent study or revision of the given species, but with the ever-increasing tendency toward better care and preservation of types in the museums this is not so serious, and if "It is the duty of the next investigator to select one among the syntypes as a lectotype" (Spath, 1934, p. 45), the responsibility is also his.

OUTLINE OF PALEOGEOGRAPHY

In any outline of regional paleogeography, first consideration is to be given to the index fossils—"The sole international medium"[17] in geology which are as indispensable for a broader interpretation of the principal seaways and the location of their paths of advance as they are for the purpose of stratigraphic correlations.

In what direction was the communication route of the parahoplitan and acanthohoplitan faunas found in the depositional areas of southeastern Arizona? It is obvious that these ammonites of so strong a Caucasian-Transcaspian relationship could not have migrated through the Texas and Coahuila-Durango regions of the Mexican geosyncline, since no conspecific or even closely allied forms are known in the strata deposited in the Mexican sea. It has been argued in discussions with colleagues engaged in stratigraphic research in Texas that such a fauna as, for instance, the parahoplitan fauna of the Pacheta member may be buried in the Las Vigas formation in which it probably remains undiscovered, as was the ammonite fauna of Bisbee. This is a possibility, but too large areas of Texas and northern Mexico have been covered by sufficiently extensive geological survey without encountering similar faunas to admit this conjecture as a working theory.

There are no faunal successions in the Mexican geosyncline comparable to that of Bisbee. According to Kellum (1936b, p. 1066-1071), in the Laguna District of southwestern Coahuila and eastern Durango the beds with *Neocomites? praeneocomiensis* are succeeded by strata with a lamellibranch fauna of the Malone facies, and the entire sequence terminates abruptly with the *Dufrenoya* beds. Note the complete absence of a parahoplitan fauna below the *Dufrenoya* zone and of an acanthohoplitan fauna above it. This is in agreement (1) with the development of lagoonal gypsiferous deposits on the west side of the Mexican Aptian sea in Coahuila and Durango (and probably also in Chihuahua) which, either independently or in conjunction with other paleogeographic factors, had isolated on the southeast side the recurring open sea facies of Arizona (Burckhardt, 1930, p. 163, fig. 47 and p.

---

[17] Arkell, 1933, p. 2.

212, fig. 55; Schuchert, 1935, p. 171, fig. 26; Kellum, 1936a, p. 424, fig. 2); and (2) with the absence of ammonites related to the Parahoplitinae and Acanthohoplitinae of Arizona in the northern half of the Mexican geosyncline to the northwest, and for a long distance to the southeast. Either a straight or circuitous intercommunication may have existed, however, in *Dufrenoya* time (the dufrenoyan fauna and the Aptian Trigoniae of the Malone facies are common to both southeastern Arizona and southwestern Coahuila—eastern Durango).

On the more positve side are: (1) Fragmentary specimens from the Mazapil area, Zacatecas, described by Burckhardt (1906, p. 191–198, Pl. 42, figs. 8–10; Pl. 43, figs. 1–11). It is rather difficult to evaluate, from this assemblage, how close Burckhardt was in his comparison of some of these incomplete examples of ammonites with the Upper Aptian parahoplitids of the Caucasus and France, but there is an unmistakable resemblance. (2) Fragmental material from the San Carlos Mountains, Tamaulipas, described by Kellum (1937, p. 76–82, Pls. 8–9), an ammonite specimen of which was compared to a Caucasian species. Böse (1927, p. 60) mentioned "*Parahoplites*" *aschiltaensis* and *Parahoplites* aff. *melchioris* from the San Carlos Mountains. To this may be added "*Parahoplites*" cf. *multispinatus* and "*Uhligella*" aff. *walleranti* listed by Kellum (1937, p. 32; Tables 1, 2). Since no satisfactorily preserved examples of these materials have been illustrated or described, it is impossible to determine the relation of these interesting ammonites to the species from the Caucasus and Arizona. However, evidence is insufficient to suppose that this fauna migrated northwestward toward Texas. (3) Undescribed specimens of *Acanthohoplites* and *Beudanticeras* observed by Imlay (1939a, p. 1734) in the El Tigre area of northeastern Sonora. This collecting locality, however, is close to the international border south of Bisbee. The horizon from which these specimens were cited most probably corresponds with the Quajote member of my description. (4) The fauna of Rio Nazas, Durango (Burckhardt, 1925). In a preliminary account of the Cretaceous stratigraphy of southeastern Arizona (Stoyanow, 1938, p. 117), a relation between the ammonites of Rio Nazas and Bisbee was maintained. Further studies of the Arizona ammonite faunas have made this contention untenable. According to Burckhardt (1925, table opposite p. 54; 1930, p. 140, fig. 38), the Rio Nazas fauna is partly in beds with *Dufrenoya* and partly above them. I believe that: (a) this fauna is older than the acanthohoplitan fauna of the Quajote member (D. 3d-3b); (b) its stratigraphic position is with and immediately above the *Dufrenoya justinae* zone; and (c) its place in the Bisbee sequence most probably is in the upper part of the Cholla member and the lower part of the Quajote member (D. 5f—D. 3e).

The relation of ammonites from Rio Nazas to the European and Eurasian species was summed up by Burckhardt (1925, opposite p. 44; *see also* Burckhardt, 1930, p. 141). This fauna needs revision. It appears to be close to the asemblage reported by Böse and Cavins (1927, p. 19; Burckhardt, 1930, p. 143) from the Cañon de Vallas near the Hacienda Saucillo northeast of Saltillo, Coahuila, where the dufrenoyan, cheloniceran, and acanthohoplitan faunas seem to be present in close succession. The location of the Rio Nazas fauna, considerably northwest of Mazapil, is of interest. If the locations of Mazapil, Rio Nazas, El Tigre, and Bisbee are connected on the map with a pencil line, the line will be close to the western border of the Mexican sea and west of the Aptian lagoonal deposits established in Durango and Coahuila by Kellum (Burckhardt, 1930, p. 163, fig. 47, Schuchert, 1935, p. 162, fig. 25; Kellum, 1936a, p. 425); it also will connect with the bathyal region of the Mexican geosyncline which is supposed to have opened in easterly direction (Schuchert, 1935, p. 171, fig. 26).

Yet, the possibility of ammonite intercommunication along the western border of the Mexican sea from the Atlantic region cannot explain the strong Caucasian-Transcaspian affinities in the parahoplitan and acanthohoplitan faunas of Bisbee. As has been pointed out, there are no comparable ammonites in the geosyncline except a very indefinite suggestion at Mazapil and, nearer the Atlantic border, in the San Carlos Mountains. It is also rather surprising that according to Burckhardt (1906, p. 191–198), Böse (1927, p. 21, p. 60), and Kellum (1937, p. 79) the specimens identified as congeneric with *Parahoplites* and *Acanthohoplites* are referable not to the Gargasian and Clansayan species of western Europe, as should be expected if these forms were of the Tethyan-Atlantic derivation, but directly to the species of Mangyshlak and the Caucasus. The strata with *Acanthohoplites* and *Beudanticeras* of the El Tigre area in northeastern Sonora, which have been mentioned by Imlay, most probably belong to the depositional area of Bisbee and are a part of the Lowell formation. Imlay (1939a, p. 1735) noted the absence of such faunas in the collections from the strata associated with the *Dufrenoya justinae* zone of north-central Mexico. Similarly, in the interesting sequence at Arivechi, east-central Sonora, there are no species comparable to the ammonites and Trigoniae of southeastern Arizona (Burckhardt, 1930, p. 176; King, 1939, p. 1670; Imlay, 1944, p. 1036).

The inviting alternative is that the communication path of the Arizona faunas under discussion was in a western direction. It may be argued that in Aptian time there was a land mass in central Sonora which partially limited the Mexican geosynclyne on the west, as has been summed up by Imlay (1939b, p. 425), and that no ammonites comparable to the parahoplitids of Arizona are known west of the El Tigre area in Sonora and in Lower California.[18] Conflicting with this, however, is direct evidence showing the presence of marine Aptian strata in northwestern Sonora. Burckhardt (1930, p. 150) has reported Trigoniae of the pseudo-quadrate and *aliformis* groups near Santa Ana in "central" Sonora[19] which he compared with South African species and which most probably are more related to the Trigoniae from the Lowell formation. It will be shown that there are certain species in the Lowell formation which could have entered the Bisbee area only through northwestern Sonora.

Imlay (1939a, p. 1739) postulated that in Trinity time land in northeastern Sonora

---

[18] The strata with rudistids and caprinids in Lower California (*fide* Burckhardt, 1930, p. 209) obviously are much younger.

[19] This is a confusing error. In the text Burckhardt discussed this locality as being in central Sonora, but in his orientation map (Burckhardt, 1930, p. 163, fig. 47) he showed the location of Santa Ana near Magdalena, which is almost due south of Nogales, Arizona, and only 45 miles from the international border. Since there is another Santa Ana, which is much closer to the geographical center of Sonora (near Tonichi, approximately at 29.00° N. Lat. and 109.00° W. Long.), I have queried Dr. Theodoro Flores, the original finder of the fossiliferous locality in question. Dr. Flores very kindly informs me that the fossils mentioned by Burckhardt were collected at Santa Ana near Magdalena. The strata of Cambrian, Mississippian, Pennsylvanian, Permian, Triassic, and Jurassic ages are already known in the Magdalena-Altar Valley. There are weighty reasons to consider the Altar Valley as the possible portal to the west during several geological periods.

and southeastern Arizona limited the Mexican sea on the west, thus eliminating the possibility of a western embayment between El Tigre and the international border, but that a western arm from the sea on the north side of that land extended probably from Bisbee to the Patagonia Mountains.

As a mater of fact there is no depositional continuity between the Bisbee area and the Patagonia Mountains on the territory of Arizona. On the evidence of accumulated observations, I cannot regard the Lower Cretaceous paleogeography of southeastern Arizona as a single phase determined by a uniform depositional development in a restricted and simple basin of a constant general outline. I have already mentioned oscillatory conditions and several recurrences of the open sea facies during the deposition of the Lowell formation. I will attempt now to show by paleontological and lithologic data that the configurations and connections of the Lower Cretaceous basins in Arizona changed several times.

The Lowell formation and the Mural limestone have not been observed west of the Bisbee area. A brief description of the Cretaceous outliers to the west of Bisbee, viz., (1) south of Tombstone, (2) on the westerly side of the Huachuca Mountains, and (3) in the Patagonia Mountains, has been presented. There is no evidence that the Cretaceous strata of these isolated areas were formed in a single depositional basin. On the contrary, there is strong evidence that they are of different ages: (1) The lamellibranch fauna of Tombstone collected so far does not permit a comparison with any faunal assemblage of the Bisbee area, but the lithologic development of the strata suggests their post-Mural age. (2) The examined sequence on the west side of the Huachuca Mountains is different lithologically from that of Bisbee, but these strata contain Glen Rose fossils which, however, have not been observed either in the upper part of the Lowell formation or in the Mural limestone. Imlay (1944, p. 1037) mentions the presence of *Exogyra arietina* high on the top of the Huachuca Mountains, but he neither refers to the lithologic nature of the stratum involved and its position in the section, nor states whether or not it rests on the Mural limestone or its equivalent. (3) The Patagonia group of the Patagonia Mountains, a sequence composed mainly of dark shales sporadically interbedded with arenaceous strata and volcanics, and in the upper third with bedded arenaceous limestones that contain a prolific fauna of *Stoliczkaia* (Molly Gibson formation), is 4500–5500 feet thick and is lithologically different from the strata of the areas to the east. The lower part of this group, below the *Stoliczkaia* zone, may be younger than the Mural limestone and the Cintura formation, and represent in part or entirely the Fredericksburg and the Washita of Texas, or equivalents of the older stratigraphic units of the Bisbee group may also be present in the lowest strata, but quite evidently the depositional basin of the Patagonia group has never been directly connected with the Bisbee area.

In 1942 I postulated the Altar Headland as the Mexican part of the pre-Cambrian granitic-schistose platform—Mazatzal Land of Arizona—which separated the Paleozoic depositional areas of Arizona and northwestern Sonora (Stoyanow, 1942, p. 1264). There is evidence that the western Sonoran trough still existed, probably with interruptions, throughout the Mesozoic era, during which time the land mass was subject to erosion, especially in Cretaceous time. If it can be shown that most of the marine Cretaceous fauna of southeastern Arizona had only the western communication open through the Altar Valley in Sonora, the discussed depositional areas may be regarded as a result of isolated periodical marine incursions from the south which were controlled by the rapidly changing land and sea relations. The times when the depositional basins were connected with the seas in Texas, either directly or through northern Mexico, are clearly indicated by the presence of common species at certain horizons of the general sequence.

Parahoplitidan faunas are known from Shasta County, California. In his monograph on the Lower Cretaceous deposits and faunas of California and Oregon, F. M. Anderson (1938, p. 76-90) discussed at length the faunal contrast of the Indo-Pacific and the Tethyan-Atlantic provinces. Pointing out that the lithologic and faunal differences between the Comanche and Shasta series had been known since Stanton's (1897) research and that there are no identical ammonite species in the Cretaceous strata of Texas and California, Anderson made an important contribution by demonstrating the affinity between the parahoplitidan faunas of California, on the one hand, and of Mangyshlak and the Caucasus on the other. Very unfortunately Anderson's descriptions and illustrations of ammonites were made almost exclusively from larger shells without an adequate examination of the inner whorls. Also the lack of figures showing sutures, ventral views, and whorl sections considerably handicaps correct interpretation of described species. As a consequence, there may be serious misinterpretations of generic relations.[20] On the basis of the presented material no species of the California assemblage can be connected directly with the species and varieties established by Anthula, Sinzow, and Kazansky from the Caucasus and Mangyshlak since the comparison was made on the approximate resemblance of larger whorls and in the case of the forms identified as *Acanthohoplites* the presence of tubercles was not demonstrated in any of the described specimens. The general characteristics of the fauna are very similar, however, and a study of the inner whorls and of the individual development may reveal a closer relationship.

The Parahoplitinae and the Acanthohoplitinae of southeastern Arizona, on the other hand, are easily traceable to the species of the Caucasian and Transcaspian realm. Such species as *Kazanskyella arizonica*, *Paracanthohoplites meridionalis*, *Acanthohoplites impetrabilis*, *A. teres*, and *Immunitoceras immunitum* are directly comparable to the described Russian species. In some other species, however, such as *Acanthohoplites schucherti*, *A. berkeyi*, and *A. hesper*, the affinity is less definite.

Emphasizing the similarity between the Lower Cretaceous ammonites of California and the Caucasian region, Anderson (1938, p. 80-86) discussed the hypothesis of a late Mesozoic land bridge across the Pacific Ocean and brought out the views on the problem which had been postulated by Burckhardt, Schuchert, Gregory, and others. Favoring the southern route of communication, he seemed to be somewhat puzzled by the remoteness of the Caucasus. In this connection I may briefly sum up the progress of our knowledge of the Upper Aptian parahoplitids and their geo-

---

[20] For instance, the trifid lateral lobe in the suture of *Parahoplites dallasi* Anderson (1938, Pl. 34, fig. 2) is too symmetrical for this genus (*compare* with the typical parahoplitan sutures in Plate 17, figures 9 and 10). *Parahoplites stantoni* Anderson (1938, p. 169) is compared to "*Parahoplites*" *uhligi* Anthula which most probably belongs in the Acanthohoplitinae. On the other hand, the suture of *Acanthohoplites aegis* Anderson (1938, p. 171, Pl.36, fig. 1), which has a characteristically acanthohoplitan first lateral lobe and most probably is a true acanthohoplite, is regarded as similar to the suture of *Parahoplites*.

graphic distribution. When Anthula (1899) described the parahoplitids of Daghestan it appeared to be a more or less localized fauna. Further interest in it was stimulated by Jacob's (1905) research on the Clansayan paleontology and stratigraphy, and by Stolley's (1908a) stratigraphic interpretations. Sinzow's monograph on the faunas of Mangyshlak and the Caucasus appeared in 1908 but did not attract much attention until its bearing on the stratigraphy of western continental Europe (Kilian, 1913) and England (Spath, 1930) was fully realized. Kazansky's (1914) work added many new species and varieties to the parahoplitids of the Caucasus but unfortunately is not known widely because it was published only in Russian. Renngarten (1931) gave a comprehensive account of the stratigraphic distribution of the Cretaceous faunas, including the parahoplitids, in the Caucasus. There is some evidence of the southward extension of the areas with the parahoplitan faunas of the Caucasus and Mangyshlak: in Luristan (H. Douvillé, Mission scientifique en Perse par J. de Morgan, t. 3, pt. 4, Pl. 28, figs. 2a–2b, Paléontologie, 1904, Paris) and probably also in Khorasan (Bogdanowitch, 1890, p. 131, fig. 13; Pl. 3, fig. 5). Anderson's studies in California and mine in Arizona show the presence of related faunas in the southwestern part of the United States.

It is true that no closely allied faunas are yet known in central and eastern Asia. But undoubtedly intermediate areas, now nearly all in the undeveloped countries that still await advanced exploration, will be established with the further progress of international geology. There are numerous examples of closely allied or even identical faunas in widely separated regions. To illustrate the long geographic distances between areas with congeneric specialized faunas, I may mention a case connected with my own research. After the publication of the papers dealing with a productoid brachiopod, *Tschernyschewia*, which was discovered in the Upper Permian strata near Djulfa on the Russian-Persian border south of Mount Ararat (Stoyanow, 1910, p. 853; 1915, p. 55, Pl. 1, figs. 1a–4d, Pl. 2, figs. 1a–11b) it was shown that the only known congeneric forms occur in the Loping area of Province Kiangsi, eastern China (Frech, 1911, p. 132. Discussed and emended by Chao, 1928, p. 75–79). The distance between Djulfa and Loping, roughly measured, is about 4500 miles. *Tschernyschewia* is a highly specialized brachiopod of a short time range, therefore it is imperative to accept faunal intercommunication between Armenia and eastern China in late Permian time.

The trend of the present discussion is corroborated by analysis of the geographic distribution of the faunas that are most closely related to the Cretaceous paleontological assemblages observed in southeastern Arizona, in which the elements of (a) Pacific-Caucasian, (b) Texas-East Mexican or Tethys-Atlantic, and (c) South American faunas are distinguished.

In the basal fossiliferous strata of the Bisbee group, that is, in the Lancha limestone (Pacheta member of the Lowell formation, D. 9g), occur numerous specimens of *Acila (Truncacila) schencki*. A glance at the world map prepared by Schenck (1936, p. 32, fig. 9) to illustrate the distribution of the Cretaceous, Tertiary, and living species of *Acila* will show the Pacific geographic relation of the Arizona species, though the present location of *A. schencki* is considerably inland. It is true that in the strata of the Pacific Coast no species of *Acila* is older than the Upper Cretaceous.

The only other known Lower Cretaceous species of *Acila*, *A. (Truncacila) bivirgata* (J. de C. Sowerby), the type locality of which is in southeastern England, is said to occur also in northwestern continental Europe, northwestern Africa, and Venezuela. Schenck (1936, p. 40) suggests possible communication between Europe and northern South America. However, no specimens of *Acila* have been recorded from the Cretaceous deposits of Mexico and Texas. Plotting the location of *A. schencki* on Schenck's map leaves no choice but to associate it with the Pacific province.

No less suggestive of western connections are the ammonites which in the Lowell formation are below the strata with dufrenoyan faunas: *Kazanskyella arizonica* (Pacheta member, D. 9c) and *Paracanthohoplites meridionalis* (Joserita member, D. 8e). The first ammonite, *Kazanskyella arizonica*, is very close to the Caucasian *K. daghestanica* (nom. nov.); both these species show a definite relation to *Parahoplites melchioris* Anthula i.e., to the true *Parahoplites*. On the other hand, the species of western Europe, eastern Mexico, and Texas which have been referred to *Parahoplites* deviate considerably both from the genotype and from several species described by Sinzow from Mangyshlak. Quite possibly the scanty parahoplitan fauna of Mazapil, Mexico, indicates an infiltration from western stations similar to that of Arizona. The species referable to *Parahoplites* which have been described by Gayle Scott (1940, p. 978, 1029–1033) from the Blue Marls of the Cuchillo formation in the Quitman Mountains of Texas are of post-*Dufrenoya* time. The second ammonite, *Paracanthohoplites meridionalis*, is related to *P. multispinatus* (Anthula) from the Caucasus.

The post-dufrenoyan fauna of the Acanthohoplitinae in the Quajote member of the Lowell formation has many features in common with the assemblages described by Sinzow (1908) from Mangyshlak, notwithstanding which, and like the latter, it is essentially an autochthonous fauna.[21]

The South American element is sparsely represented in the Lowell formation, preeminently by the Trigoniae of the *abrupta* and *excentrica* groups, and by the pseudoquadrate Trigoniae of the Malone facies.

There is a number of lamellibranch species in the Lowell formation which are exotic and unknown, thus far, elsewhere in the Cretaceous deposits of North America. Some are suggestive of European species, which, however, have not been observed in the Mexican geosyncline, whereas others also show a relation to certain species of northwestern South America. These are: *Exogyra lancha*, *Gervillia cholla*, *G. heinemani*, *G. rasori*, *Idonearca stephensoni*, *Lima cholla*, *L. espinal*, *Ostrea edwilsoni*, *Pecten (Chlamys) thompsoni*, and *Pteria peregrina*.

The infiltration of Texan species in the Bisbee basin was very gradual but increased markedly toward the final stage of deposition of the fossiliferous strata. In the Lowell formation no specimens unquestionably identical with the described lamellibranch species of Texas have been collected below and in the beds with *Dufrenoya*. But above the *Acanthohoplites* zone (Quajote member), in the upper part of the for-

---

[21] I use the term "autochthonous" not in Buckman's sense, *i.e.*, "abundant and well-preserved ammonites" as distinct from those which drifted away from the regions where they have been autochthonous (*fide* Arkell, 1933, p. 562), but for the prolific faunas which were capable of producing many modifications of specific and varietal degree within a relatively limited environment.

mation, the Perilla member has yielded in abundance such fossils as *Trigonia mearnsi* and *T. stolleyi* which, with *Orbitolina texana* in the basal beds of the Mural limestone, clearly indicate communication with the depositional areas of Texas in late Cuchillo and early Glen Rose time. The earlier communication of *Dufrenoya* time may have been indirect, through the Mexican part of the sea. Was there a direct connection between southeastern Arizona and Texas in the time of deposition of the Perilla and the Mural?

The Mural limestone is the most stable stratigraphic unit of the Cretaceous sequence in southeastern Arizona. It is of the same lithologic composition and contains the same fauna in all observed sections, which is quite different from the striking localization of the earlier (Lowell formation) and the later (Molly Gibson formation) facial developments. The observer traveling along the international border from Bisbee to Douglas sees the Mural limestone not only in the outcrops along the highway on the American side, but also on Mexican territory where the Mural limestone appears as a cap rock in the tilted and parallel fault blocks. This continues through Douglas to the southeastern corner of Arizona where the Guadalupe Mountains cut across the border east of San Bernardino (*compare* Burckhardt, 1930, p. 178; p. 212, fig. 55). Darton (1928, p. 38–39) mentioned that in southwestern New Mexico the limestone beds in the Hatchet Mountains region closely resemble the Mural limestone of Bisbee and Douglas, and that similar limestones with capprinids and Trigoniae also occur in the Potrillo Mountains east of El Paso. But Lasky (1938, p. 530, fig. 4) has shown that in the Little Hatchet Mountains the equivalent of the Glen Rose is 15000 feet thick and that the beds with *Orbitolina* and rudistids recur three times above the strata with *Trinitoceras* and *Douvilleiceras* (Gayle Scott, 1940, p. 993, 1019, 1015), whereas in between occur volcanic rocks and freshwater deposits. On my visit with Dr. Scott to the Quitman Mountains, north of the international border near Indian Hot Springs, Texas, I was impressed by the great thickness of the limestone sequence with *Orbitolina texana* above the zone of *Trigonia mearnsi* (with *Trinitoceras rex*). It is apparent that in the sections east of Arizona the equivalent of the Mural limestone thickens considerably. Our knowledge of stratigraphic paleontology in this part of the Cretaceous sequence is rather inadequate, but there seems to be a definite uniformity in the faunal succession in all studied sections: The strata with *T. mearnsi* (Bisbee; Quitman Mountains) are succeeded by those with *Trinitoceras* (Little Hatchet Mountains; Quitman Mountains) and *Douvilleiceras* (Bisbee?; Little Hatchet Mountains; Quitman Mountains), and finally, in all sections, by the beds with *Orbitolina* and rudistids.

The known Lower Cretaceous outliers north of the international border in southeastern Arizona and southwestern New Mexico, between the Patagonia Mountains and El Paso, are oriented essentially in a southeast-northwest direction, owing, most probably, to the Basin Range structure. However, the nature of such sequences as, for instance, those represented by the Patagonia group, the Bisbee group, and the one in the section described by Lasky from southwestern New Mexico, clearly indicates marine incursions from the south and does not favor a latitudinal depositional continuity along the international border.

The northern margin of the major Aptian-Albian seaways to the south, in northwestern Mexico, reached southeastern Arizona several times and at different areas. Times of Indo-Pacific and Tethyan-Atlantic connections are indicated by the paleontological content of the studied stratigraphic units.

Western communication

Eastern communication

Patagonia group

Prolific fauna of *Stoliczkaia* in Molly Gibson formation is most probably of Indo-Pacific connection. Although several species of *Stoliczkaia* are known in Texas and in northeastern Coahuila west of the international border, there is no evidence of this fauna in the Mexican trough.

Bisbee group, Mural limestone

The abrupt western termination of Mural limestone at Bisbee and abundance of *Orbitolina texana* at several horizons indicate an eastern connection.

Lowell formation, Perilla member

Trigonia mearnsi and *T. stolleyi* are abundantly represented below the beds with *Orbitolina* in the Quitman Mountains, Texas.

Lowell formation, Quajote member

Except northeastern part of Sonora, the acanthohoplitan fauna of this member is unknown elsewhere in Mexico and Texas.

Lowell formation, Cholla member

Trigoniae of the Malone facies and the *Dufrenoya* fauna indicate a connection with Coahuila.

Lowell formation, Saavedra member

Lamellibranch fauna of this member has no counterpart in the Mexican geosyncline. *Trigonia reesidei*, allied to South American species, may be represented in Texas by undescribed forms of the *abrupta* group.

Lowell formation, Joserita member

Lower part of this member is characterized by the presence of *Paracanthohoplites*, whereas a dufrenoyan fauna is introduced in its upper part.

Lowell formation, Pacheta member

Parahoplitan and nuculid faunas of this member indicate a western connection.

APPENDIX.—NOTE ON LATE CRETACEOUS STRATA IN SOUTHEASTERN ARIZONA

There is no continuity between the discussed marine late Lower Cretaceous Molly Gibson formation of the Patagonia group in the Patagonia Mountains (Pl. 2, 7) and the sequence of fresh-water Upper Cretaceous deposits exposed along the eastern front of the Santa Rita Mountains (Pl. 2, 8).

On Schrader's geological map of the Santa Rita and Patagonia Mountains, the eastern side of the Santa Rita Mountains is shown as composed of Quaternary gravel and sand. Schrader (1915, Pl. 2; p. 51–56) mapped in this region as Cretaceous only a small and isolated area in the head of Adobe Canyon in which he distinguished between the stratigraphically lower "Mesozoic rocks," about 600 feet thick, and the overlying fresh-water strata with *Unio* and *Viviparus*, supposedly of Eocene age, as identified by Stanton.

Examination of several parallel canyons of a general southeasterly trend, which cut across the eastern slope of the Santa Rita Mountains between Patagonia and the Greaterville fault, shows that the nearly horizontal sub-Recent and Recent terraces made up of gravel, sand, and volcanic ash extend from the Sonoita Flat only to the lower parts of the canyons, where they rest on the easternmost outcrops of the upturned Cretaceous strata here discussed. The upper parts of the canyons and the intervening divides are formed of a continuous sequence of tilted and often almost vertical Cretaceous strata, apparently several thousand feet in total thickness, which uniformly strike northwest and have a northeastern regional dip.

Immediately north of Patagonia are multicolored and bedded soft volcanics. Their relation to the sequence is not quite clear, but farther north, in the Casa Blanca Canyon, stratified hard andesitic flows and breccias are overlain with apparent conformity by the great series of sedimentary strata named Sonoita group in one of my preliminary reports (Stoyanow, 1936, p. 296).

The Sonoita group readily admits of two divisions: the lower—Fort Buchanan formation, and the upper—Fort Crittenden formation, both names adopted from the near-by ruins of two old forts.

The Fort Buchanan formation rests on the andesitic flows with a basal conglomerate about 400 feet thick. Above the conglomerate, a series of strata, approximately 1500 feet thick, is composed of alternating hard gray sandstones and soft maroon shales. Because of the rapid weathering of the shale, the vertical, wall-like beds of the hard sandstone form "ruins", a unique landscape feature for southeastern Arizona. Schrader referred to the upper part of the Fort Buchanan formation as the "Mesozoic rocks" of the eastern side of the Santa Rita Mountains.

The Fort Crittenden formation is separated from the Fort Buchanan formation by a comparatively thin conglomeratic bed, but it is composed of markedly different strata. In ascending order above the conglomerate are: (1) Alternating yellow thinner-bedded shales and hard sandstones, 600 feet thick, with numerous specimens of *Unio, Viviparus, Physa*, and remains of fishes and turtles. Like Schrader, I was inclined to regard these strata as Tertiary. However, a find of dinosaurian teeth and bones clearly indicates a pre-Eocene age for this unit. I am indebted for this identification to Dr. Barnum Brown (personal communication, October 16, 1941) whose kind answer to my query reads as follows:

"In reply to your letter of Sept. 22nd, the two dinosaur teeth—DC 55 and DC 56—came in due time and after comparing them with specimens in our collection I can say definitely that they are from a carnivorous dinosaur skull of about the size of a large specimen of *Gorgosaurus libratus*—from a skull probably 2-½ to 3 feet in length."

(2) 200 feet of yellow, red, and black shale alternating with hard yellow and pink sandstone. These beds contain the same invertebrate fauna as in the underlying unit. (3) Light-colored conglomerate with ingredients rapidly changing in size and composition, 350 feet thick. (4) Thin-bedded sandstone, 80 feet thick. (5) Soft light-colored shale overlain by very fossiliferous gray and buff sandstone and shale with *Unio, Sphaerium*, and fish plates. (6) Conglomerate, probably over 1000 feet thick, composed of large boulders. (7) South of the Greaterville fault the sequence terminates in a thick series of scarlet red strata.

The invertebrate fauna of the unit 1 also occurs in the southwestern foothills of the Whetstone Mountains, near the northeastern margin of the Sonoita Flat.

The Sonoita group strongly resembles the Cabullona group described by Taliaferro (1933, p. 12–37) from northeastern Sonora, southwest of Douglas, Arizona, which rests unconformably on Lower Cretaceous strata and in which also occur the fresh-water mollusks and dinosaurs.

Fossil evidence is very scarce in the larger area to the northeast between the Greaterville fault and the Empire Mountains, mapped by Schrader as Cretaceous, which differs from the discussed area in abundance of volcanic material and in the absence of clearly outlined stratigraphic units. Poorly preserved lamellibranchs were collected in limestone near Cuprite Mine in the terminal northern foothills of the Santa Rita Mountains, and northeast of the Greaterville fault, in clastic rocks near Martinez Ranch and in the Fish Canyon. The examined fragmental material does not suggest a fresh-water fauna and probably belongs to a brackish **facies.**

CHAPTER 2, PALEONTOLOGICAL PART

PELECYPODA

Family NUCULIDAE d'Orbigny
Acila H. and A. Adams, 1858
Acila (Truncacila) schencki Stoyanow, sp. nov.

(Plate 8, figures 1–8)

MEASUREMENTS:

Syntype	No. 91029	No. 91030	No. 91053
Length	14	14.5	
Height	10.5	11	10

EXTERIOR: Shell small, 14 mm. of average length, longer than high, trigonal-oval in outline, truncate posteriorly and narrow anteriorly; antero-dorsal margin, almost a straight line except in vicinity of umbones and of anterior extremity, steeply inclined forward; anterior extremity narrowly rounded; ventral margin smooth, moderately convex, evenly curved from anterior to posterior margin with greatest convexity anterior to umbones; posterior margin straight, forming right angle with the tangent to dorsal margin; umbones blunt and rounded; beak opisthogyral, pointed; escutcheon shallow, ill-defined; lunule small, shallow.

Ornamentation consists of closely set ribs arranged in chevron-like inverted V-s. The current terminology of sculpture in *Acila* is somewhat vague. Schenck (1936, p. 16), referring to the ribs of *Acila* as "radial," evaluated the terms "bifurcation" and "divarication" as applied to this kind of ornamentation with a preference for the latter. Bifurcation, as Schenck stated, presupposes splitting of the ribs of a higher order, and indeed in the morphology of brachiopods is often opposed to the appearance of new ribs by intercalation. However, divarication means a forking or stretching apart of something that previously was not forked. This term, used by Schenck only to characterize a definite ornamentation, in itself may be construed to imply the necessity of postulating the origin of chevron-like ribs out of simple ribs of some ancestral forms through separation. Quenstedt (1930, p. 81 *and* "Erklärung der Tafel 2"), who advocated a mechanical theory of ribbing in Nuculidae, distinguished between the radial ribs (as in *Nucula pectinata*) and the divaricate ribs (as in *Acila bivirgata*). It seems that the application of terms like V-s, inverted V-s, chevrons, zigzagging ("secondary divarication" of Schenck), successfully employed by Kitchin and others in similar cases, is simpler and more to the point, inasmuch as Woods (1903, p. 19) has shown that in *Acila* (*Truncacila*) the new ribs may appear both by bifurcation and by intercalation. Besides, this eliminates the argument in describing the curved chevron-like ribs as "radial," which they are not, since each pair issues at an individual point independently of the rest of the set.

The line bisecting the chevrons, or Schenck's line of primary bifurcation, is slightly anterior to the greatest convexity of umbones and anterior to the center of the shell. This allows an easy separation of the anterior from the posterior ribs. It is in the posterior ribs of the dorsal region, also down the flank in close vicinity of the escutcheon, and sometimes in the anterior ribs in the region of antero-ventral extremity, that a slight zigzagging or "secondary divarication" is observed. This often is so minute as to be easily missed by the naked eye.

There is a weak ridge which separates the shallow escutcheon from the flank, but there is no depression, sulcus, or groove to separate the ridge from the central part of the escutcheon; also there is no ribless area in the escutcheon and the ribs pass uninterrupted from the flank directly onto the escutcheon.

In some specimens the concentric lines of growth occur only in the vicinity of ventral margin; in others they also are present high on midflank, and sometimes even in umbonal region. In no part of the flank do they obscure or eradicate the ribbing.

Truncacilae lack the rostral sinus of Acilae proper, as has been pointed out by Schenck. In many Truncacilae a rostration at the point of meeting of the ventral and posterior margins is quite well developed, and in some there is a shallow embayment in the ventral margin just anterior to the rostration, this may be referred to as *"rostral dent"* (*compare* Schenck, 1936, Pl. 5, figs. 8, 11). In *Acila (Truncacila) schencki* the rostration and the rostral dent are altogether absent. Both *Acila (Truncaila) bivirgata* (J. de C. Sowerby) and *Acila (Truncacila) demessa* Finlay, two species of Cretaceous age, possess a strongly developed and well-outlined rostration (Woods, 1903, Pl. 3, figs. 1–12; Schenck, 1936, Pl. 2, figs. 1–15). Recently described Late Cretaceous species, *Acila (Truncacila) princeps* Schenck (1943, p. 60, Pl. 8, figs. 1–4, 6–8), differs from *A. demessa* "chiefly in its larger size and coarser ornamentation." Senonian *Acila (Truncacila) hokkaidoensis* (Nagao, 1932, p. 28, Pl. 5, figs. 17–18) is "provided with a somewhat produced and subangulated end," i.e., the rostration. Rostration seems to be a constant character in Eocene *Acila (Truncacila) decisa* (Conrad, *in* Schenck, 1936, Pl. 3, figs. 1–9, 11–15) and in Oligocene *Acila (Truncacila) shumardi* (Dall, *in* Schenck, 1936, Pl. 6, figs. 1–11). In Miocene *Acila (Truncacila) conradi* (Meek) it is less pronounced and apparently lacks in certain specimens (Schenck, 1936, Pl. 8, figs. 1–3, 5–10, 12–14). In Pleiocene-Recent *Acila (Truncacila) castrensis* (Hinds) this feature, when present, is barely indicated angulation between ventral and posterior margins (Schenck, 1936, Pl. 10, figs. 1–15). In the general outline *A. schencki* resembles species with weak rostration and narrower anterior extremity, as in certain specimens of *A. conradi* figured by Schenck (1936, Pl. 8, figs. 5, 8, 9, and 12), and is quite distinct from Cretaceous *A. bivirgata* and *A. demessa* with a strong rostration and a broadly outlined anterior part of the shell.

INTERIOR: In the hinge of *A. schencki* there are 18 or 19 anterior and about 7 posterior teeth. The early anterior teeth of right valve, in an appreciably deep depression, are markedly obsolete but always distinct. In left valve the early teeth of anterior dentition are set on a projecting ridge which fits exactly into the depression of the anterior row of teeth in right valve.

In right valve the flange beneath the umbo, and between the anterior and posterior sets of teeth, is more elevated and vaulted upward in its posterior part. This part is a spoon-shaped chondrophore-like platform which bears a narrow and deep pit (Pl. 8, fig. 6). The latter extends under the beak toward the escutcheon from which it is separated by a low callosity placed normal to the axis of the pit. The resilium-pit is a trigonal depression excavated in the lower, anterior, part of the flange between the buttress of the spoon-shaped platform and the anterior row of teeth. It is not placed on a distinct and individualized resilifer or chondrophore.

According to Quenstedt's (1930, Pl. 2, fig. 11) interpretation of the flange in *Nucula pectinata* Lamarck the chondrophore-like platform of this description corresponds to the "Grube für den Bandgrubenzahn," and the resilium-pit to the "Bandgrube" of right valve. I have prepared, not without difficuty, the hinge of a left valve of *A. schencki*. So far as I can see from this less perfect specimen, it hls in its flange a similar trigonal excavation opposite the resilium-pit of right valve. But instead oa a "Bandgrubenzahn" of Quenstedt there is a shallow depression for reception of the chondrophoref-like platform. Instead of a tooth-like projection to fit in the pit of that platform, there is a corresponding canal which likewise is cut off from the eschutcheon by a low callosity.

Following Quenstedt's (1930, p. 46–47) account of the development of "Bandgrubenzahn" it appears that it is an unstable formation in the hinge of Nuculidae and may occur in either of the valves. He also refers to the observation of Cerulli-Irelli that the left valve of *Nucula placentina* may develop a cavity instead of a tooth. Further, he lays stress on the fact that not all the living Nuculidae possess that tooth. It seems indeed more plausible that in certain nuculids similar plastic modifications at the proximal end of the posterior row of teeth may be associated with the ligamental development rather than represent incipient cardinal teeth (Cerulli-Irelli, *fide* Quenstedt, 1930, p. 46–47) in a well established taxodont dentition.

In general, the hinge of *Acila schencki* is developed more as in certain species of *Nucula*, such as *Nucula (Nucula) capayensis* Schenck (1939, Pl. 6, figs. 18 and 20), in which the flange is relatively wide and distinctly separates the two rows of teeth, and the resilium-pit is only partly overlapped by the anterior set of teeth, than as in a typical *Acila* (Schenck, 1936, p. 13, fig. 5, sketch 6), in which the anterior row of teeth almost overlaps the posterior row over a strongly projecting chondrophore. Compared to the hinge in right valve of *Nucula (Nucula) nucleus* (Linne), the chondrophore-like platform of *A. schencki* would correspond to the chondrophore, and the resilium-pit to the shallow trigonal space on the flange between the chondrophore and the anterior row of teeth (Schenck, 1934, Pl. 5, fig. 1a).

In *A. schencki* the adductor muscle scars, with buttresses on their inner side, are rather large for the size of the shell and are deeply impressed; the anterior scar is the largest.

FAECAL PELLETS: Three faecal pellets have been found in the Lancha limestone close to a shell of this species near its ventral margin. Compared to the pellets of *Nucula nucleus* and *Acila castrensis* (Moore and Galliher *in* Schenck, 1936, p. 11, figs. 3, 4) they are of a remarkably symmetrical cross-section which measures about 2 mm. (Pl. 8, figs. 4, 5, 7, 8). The pellets (present appearance) are composed of 6 deeply paired primary lobes, so that in all there are 12 lobes at the periphery. Distal ends of the lobes are considerably wider. Lateral view shows that the lobes extended down the length of the pellets, as in Galliher's illustrations.

If the presented interpretation of these interesting remains is correct, this is a rather remarkable case of preservation since the faecal pellets of now living Acilae are not resistant bodies. In an article on the faecal pellets Hilary B. Moore (1932, p. 307) remarked:

"It has been shown that, in the deep waters of the Clyde . . . , the simple type of pellets formed by Maldanid worms will remain more or less unchanged in the bottom deposits for periods of fifty to a hundred years, and there is no reason to suppose that they will break down after longer periods. On the contrary it is probable that they will undergo a process of mineralization."

He also refers to the observations of Takahashi and Yagi on fossilization (glauconitization) of pellets from younger Tertiary strata. On the other hand, Galliher (1931, p. 11), commenting on Moore's other observation that such pellets are "surprisingly solid and resistant; they will stand fairly rough handling, and can even be boiled with sulphuric acid or caustic soda without breaking down," says: "It seems unlikely that excrements, phosphatic or otherwise, would withstand such treatment and suggests the possibility that some other objects are being confused with the true faecal pellets." The pellets in question, probably the oldest pellets of *Acila* known to date, have been examined by R. E. Heineman, Mineralogist of the Arizona Bureau of Mines. They are composed of amorphous siliceous material with glauconite and clay minerals. A few extremely minute grains of quartz have also been observed. The pellets are pinkish-brown possibly due to the presence of iron oxide. The similarity of these bodies, both in cross-section and longitudinally, to the pellets of now living *A. castrensis* is such that there seems to be little room for mistaken identity. It should be noted, however, that these pellets, apparently belonging to a species of a smaller size, are considerably larger than those of *A. castrensis*.

TYPE: Syntypes: Nos. 91029, 91030, 91053.

OCCURRENCE: All syntypes collected from Lancha limestone, division 9g, Pacheta member of Lowell formation, Ninety One Hills.

Family PARALLELODONTIDAE Dall

Idonearca Conrad, 1862

Idonearca stephensoni Stoyanow, sp. nov.

(Plate 8, figures 9–12; Plate 9, figure 1)

Shell subquadrate in young, subquadrate-suboval in adult, slightly elongate in direction of the umbonal ridge. Anterodorsal slope, slanting toward the dorsal margin, is either flat or slightly excavated. Shallow concave posterodorsal surface, at 130° to the umbonal ridge, bears a narrow and weak elevation subconcentric to the umbonal ridge and placed closer to the latter than to the dorsal margin. Some specimens have barely perceptible broad and shallow depression in front of the umbonal ridge.

Surface is ornamented by lines of growth which are wider spaced toward the ventral margin but become closely set in its vicinity. There are more or less strong radial ribs either in anterior (Syntype, No. 92156) or posterior (Syntype, No. 91038) parts of the flank. In some specimens with strongly developed posterior ribs, are present also weak but wider anterior ribs or folds. Some specimens, however, do not develop radial ribs on the flank (Syntype No. 91831). Ligamental grooves are chevron-shaped. In hinge the subhorizontal distal teeth are at 140° to the subvertical central teeth. Other internal characters not exposed.

In general outline the ribless varieties somewhat resemble *Idonearca brevis* (d'Orbigny, *fide* Gerhardt, 1897, p. 182, Pl. 5, figs. 4a–b). Since, however, there are ribbed and ribless but otherwise identical specimens in the described assemblage, their separation on this basis does not seem justifiable.

TYPE: Syntypes: Nos. 92156, 92141, 91038, 92140, 91831.

OCCURRENCE: Lancha limestone, Pacheta member, division 9g, of Lowell formation, Ninety One Hills.

Family PERNIDAE Zittel

Gervillia Defrance, 1820

Gervillia heinemani Stoyanow, sp. nov.

(Plate 9, figures 6–7)

Holotype is a large left valve with height (measured somewhat obliquely) of about 150 mm. and length (measured along the dorsal margin) about 100 mm. This species differs essentially from *Gervillia alaeformis* (Sowerby, *in* Woods, 1913, p. 79–83, figs. 9–14) and related species in having the convex visceral elevation almost perfectly straight for two dorsal thirds. This trend is interrupted only by a very slight sinuosity in the extreme umbonal part of the valve and by a rather indistinct curve in its posterior third. Another difference is the absence of a sulcus which in *G. alaeformis* separates the visceral elevation from the posterior ear; only in the umbonal part there is a very shallow and narrow furrow that extends for 32 mm. from the dorsal margin ventrally and is much closer to the crest of the visceral elevation than to the greatest depression in the posterior ear. In connection with the lack of a prominent diagonal sulcus there is no distinct separation of the posterior ear from the visceral elevation which makes the middle of the posterior ear considerably raised with two principal slopes: one toward the dorsal margin and the other in the posteroventral direction, and with a lesser inclination toward the posterodorsal angle. The profile from the visceral elevation to the latter angle is concave. The axis of the visceral elevation forms an angle of 60° with the dorsal margin. The anterior ear is very small and passes ventrally into a concavity which corresponds with the byssal gap.

Ornamentation, as often is the case in many other species of Cretaceous *Gervillia*, is better developed in the younger, dorsal part of the shell. It is observable on the posterior ear and to a lesser degree on the adjacent part of the visceral elevation. It consists here of concentric wrinkles, somewhat coarser at intervals corresponding with projecting and stronger shell-lamellae. Radial ornamentation is represented by frequent and narrow furrows which separate the shell surface into more or less flat and wide rib-like elevations. The rest of the shell is covered only with concentric ridges which coincide with the lines of growth.

Gervillia enigma d'Orbigny (1847, p. 488, Pl. 396, figs. 9–11) seems to have a similar relation between the visceral elevation and the posterior ear, but the former considerably widens posteroventrally, and the anterior ear is much larger than in Arizona species.

TYPE: Holotype: No. 91900.

OCCURRENCE: Yellow limestone in Saavedra member, division 7i, of Lowell formation, Ninety One Hills.

Gervillia cholla Stoyanow, sp. nov.

(Plate 10, figure 1)

Shell of medium size, subrhombic in outline. Visceral elevation of the left valve, sinuous in umbonal and ventral regions, descends rather abruptly toward the anterior margin, which is concentric with its axis, and forms a more gradual and gentler slope toward the posterodorsal extremity. In the right valve (not illustrated), it is considerably narrower and drops almost vertically to the anterior margin. Beak of the left valve is more strongly developed. Anterior ear is small. In both valves the visceral elevation is not definitely separated from the posterior ear, but there is a broad and appreciably deep depression between ventral portion of the visceral elevation and the posteroventral extremity. Byssal gap is well indicated.

Following characters separate this species from *Gervillia heinemani*: subrhombic outline; sinuosity of the visceral elevation; smaller angle between the latter and the dorsal margin; absence of an appreciably concave depression between the umbonal part of the visceral elevation and the posterior ear.

TYPE: Holotype: No. 91879.

OCCURRENCE: Cholla member of Lowell formation, Ninety One Hills.

Gervillia rasori Stoyanow, sp. nov.

(Plate 10, figures 2-3)

Shell of triangular-rhombic outline with the anterior margin only slightly sinuous and oriented at 50° to the straight dorsal margin. Posteroventral margin is a single gentle curve. Visceral elevation of the left valve is wide, convex, somewhat sinuous, and not sharply separated from the sloping posterior region. Visceral elevation of the right valve is less convex, narrower, and straighter, and more distinctly separated from nearly flat posterior region. Anterior ear is small and stout. Posterior ear of the left valve in the majority of collected specimens is folded inwards.

Gervillia rasori resembles the form from Cretaceous strata of Argentina described by Burckhardt (1903, p. 70, Pl. 15, figs. 3-5) as *Perna militaris* from which it differs by a less convex and less curved visceral elevation, which is less distinctly separated from the posterior region, a more receding dorsal margin, a more concave profile of the posterior ear, and by absence of radial ornamentation. Under the same name Weaver (1931, p. 211, Pl. 15, fig. 65) described a specimen from the Quintuco formation of Argentina which seems to differ from Burckhardt's holotype in a more curved and less differentiated visceral elevation with a slenderer umbonal part and stronger radial ribs.

TYPE: Holotype (both valves): No. 91548. Paratype (left valve): No. 91958.

OCCURRENCE: Cholla member of the Lowell formation, Ninety One Hills.

Family PTERIIDAE Meek

Pteria Scopoli, 1777

Pteria peregrina Stoyanow, sp. nov.

(Plate 9, figure 5)

Only the left valve of this species is known. Its visceral elevation is but little sinuous, narrow and straight in the umbonal part, ornamented with about 17 ribs of which the most prominent, or the principal rib, occupies the most elevated position on the valve and separates the visceral elevation into two unequal parts. Posterior to this rib the visceral elevation is more rounded and descends abruptly to the posterior ear from which it is separated by narrow groove, whereas the anterior part, rounded only in the dorsal half of the valve, gently slopes toward the anteroventral region into which it gradually merges. In the dorsal half of the visceral elevation the ribs are of three sizes. The primary ribs are stronger, higher, narrower, and better rounded. The secondary ribs are lower, broader, and flatter. The tertiary ribs split off the secondary ribs. Anterior to the principal rib the sequence of ribs is: 1, a secondary rib ventrally developing a tertiary rib on its posterior side; 2, a primary rib; 3, a flat and broad secondary rib; 4, a primary rib; 5, two weak and irregular secondary ribs; 6, a primary rib; 7, two weak and irregular secondary ribs; 8, a bordering primary rib. The sequence posterior to the principal rib is: 1, a broad and flat secondary rib with a longitudinal groove; 2, a narrow primary rib; 3, a narrow secondary rib; 4, a broad and strong primary rib; 5, a narrow secondary rib, ventrally with an anterior tertiary rib; 6, a bordering rounded primary rib. Individual size of the primary, secondary, and tertiary ribs is variable. In the dorsal half of the valve the ribs are crossed by rather strong concentric wrinkles, which gives the ornamentation a cancellate appearance, whereas in the ventral region the concentric ornamentation consists only of numerous fine lines of growth. On the larger, posterior ear there are twelve irregularly alternating ribs of variable size. Strong concentric wrinkles form an imbricate ornamentation in its posterior part. The smaller anterior ear lacks radial ribbing but possesses very strong and closely set concentric wrinkles. On the whole, the radial ornamentation becomes obsolescent in the anteroventral region of the valve, where the individual length of the ribs consecutively diminishes in the direction of the anterior margin.

This species belongs in the *Pteria cottaldina* (d'Orbigny, 1847, p. 470, Pl. 389, fig. 1) group, or the first group of Aviculidae of Gillet's (1924, p. 32) classification, the representatives of which also have a single rib which is more prominent than the other ribs. The similarity is augmented by relative dimensions and order of alternation of the ribs. *Pteria perigrina* differs from D'Orbigny's species in: 1, visceral elevation of the left valve less sinuous and more attenuated in the apical part; 2, cancellate and imbricate ornamentation present only in the dorsal half of the same valve; 3, radial ornamentation absent from the anterior ear. It should be mentioned in this connection, however, that there is a discrepancy between d'Orbigny's description and his illustration of *Pteria cottaldina*. According to his text, concentric wrinkles constitute the only ornamentation of the anterior ear, whereas the illustration shows strong radial ribs crossed by fine lines of growth in that part of the shell. The angle between the principal rib and posterior part of the dorsal margin is considerably more acute in *Pteria cottaldina*. However, in *Pteria carteroni* (d'Orbigny, 1847, Pl. 390, fig. 1), which Gillet (1924, p. 32) regards only as a variety of *Pt. cottaldina*, this angle approximates that in the described species.

TYPE: Holotype: No. 91841.
OCCURRENCE: Perilla member, division 2b, of Lowell formation, Ninety One Hills.

Family OSTREIDAE Lamarck

Ostrea Linnaeus, 1758

Ostrea edwilsoni Stoyanow, sp. nov.

(Plate 11, figures 1-3)

This species varies from very narrow specimens with blunt beak (Syntype, No. 91361) to the forms considerably wider in the ventral region and with more attenuated apical part (Syntype, No. 91356). The latter variety seems closer to the types of *Ostrea praelonga* Sharpe (1850, p. 186, Pl. 20, fig. 24) from the Cretaceous of Portugal, whereas the former resembles the specimen from the Barremian of Wassy (Haute Marne), France, identified with Sharpe's species by Gillet (1922, p. 24, Pl. 2, fig. 15).

The Arizona species differs from *O. praelonga* Sharpe in a less pointed apex and tendency, in the extreme forms, toward development of a wider ventral region. It differs from *O. cortex* Conrad (1857, p. 157, Pl. 11, figs. 4a–d; Deussen, 1924, p. 39, Pl. 12, fig. 2) in the absence of strongly corrugated concentric ornamentation of that species.

TYPE: Syntypes: Nos. 91356 and 91361.
OCCURRENCE: Very abundant in Saavedra member, division 7d, of Lowell formation, Ninety One Hills.

Exogyra Say, 1820

Exogyra lancha Stoyanow, sp. nov.

(Plate 10, figures 4-6)

This species, with a shell from small to medium, elongate and crescentic in outline, falls well within the limits of the *Exogyra boussingaultii* (d'Orbigny) group. It differs essentially from the allied species like *E. boussingaultii* (d'Orbigny, 1847, p. 702, Pl. 468, figs. 6–9), *E. minos* (Coquand, 1869, p. 183, Pl. 64, figs. 1–3; Pl. 73, figs. 4–8; Pl. 74, figs. 14–15), *E. tuberculifera* (Koch and Dunker *in* Coquand, 1869, p. 189, Pl. 63, figs. 8–9; Pl. 66, figs. 12–13; Pl. 70, figs. 9–13), and *E. laevigata* Sowerby (*in* Woods, 1913, p. 404 and 406, Pl. 61, fig. 13) in that instead of the acute carina which in the left valve of these species separates the anteroventral and posterodorsal regions, there is only an obtuse angulation. This angulation is not very distinct in the umbonal part of the valve but gradually becomes stronger toward the posterior extremity, which is well seen in mature specimens. Ornamentation of the left valve is restricted to its inflated umbonal part. It consists of plications which extend from the anteroventral margin to the angulation but do not pass over it. The attachment scar is behind the spirally coiled umbo. The right valve (not illustrated), with obscure traces of ribs near the ventral margin, as in *E. minos*, is relatively strongly inflated. Some shorter, more rapidly coiling, and stronger ornamented forms found with *E. lancha* (Pl. 10, figs. 7–8) probably are closer to *E. minos* proper.

TYPE: Holotype: No. 91054. Paratypes: Nos. 91055, 91056.

OCCURRENCE: Holotype, paratypes, and specimens (illustrated) Nos. 91061 and 91063 collected from Pacheta member, division 9g, of Lowell formation, Ninety One Hills.

Family TRIGONIIDAE Lamarck

Trigonia Bruguière, 1789

I am well aware of the so-called genera or subgenera into which *Trigonia* Bruguière has been separated by some writers. The validity of certain of the introduced names undoubtedly will be substantiated by further research; others, given in a haphazard manner or ill-advisedly, are of no scientific value. Evaluation of characters of a Trigonia rendered without due regard to its ontogenic development is of but little classificatory significance. Studies show that in this very diversified branch of lamellibranchs two different species may have sufficiently similar mature shells but considerably unrelated principal characters in the young stages. Conversely, certain Trigoniae are undistinguishable in the early phases of shell development but attain very pronounced differences, including the shape of the shell, at full maturity. What Spath (1923a, p. 5) has stated relative to the ammonites, "There is a highly 'scientific' nomenclature and an extremely complex classification, but very little concrete knowledge of ammonites," may as well be applied to Trigoniae with that modification that the nomenclature of Trigoniae is established by far on a less scientific basis. A revisionary work would be far beyond the scope of this paper and ability of the writer, inasmuch as it is impossible for him to check, by the nature of descriptions and illustrations, on the ontogenic and phylogenic development of many of the "genera," the adoption or rejection of which at this time might involve an unwarranted responsibility. On the other hand, the use of "groups" and "sections" of the older paleontologists, with the full acknowledgment of the working nature of such grouping, allows a direct and unimpaired comparison between the naturally allied species without any reflection on the status of terminology of the introduced "genera" and "subgenera."

Shell terminology of **Trigonia** adopted in this paper

Inner carina separates the escutcheon from the inner area. *Median carina* is between the inner area and the outer area. *Marginal carina* separates the outer area from the flank. The term *Marginal angulation* is employed when the initial marginal carina becomes wide and obtuse posteroventrally. *Supramedian furrow* is on the dorsal side of the median carina or takes place of the latter posteriorly. *Median furrow* separates the inner area from the outer area when the median carina is not developed. *Inframarginal band* is on the anteroventral side of the marginal carina. *Marginal furrow* takes the place of the marginal carina when the latter is not developed. The term *Ribs* designates the solid or tuberculate ridges on the flank. The solid ridges, concentric or subconcentric with the lines of growth, that extend on the flank between the anteroventral margin and the marginal carina, as in some Costatae might be referred to as *Costae* in accordance with the well-established term *Costellae* which are minor solid ridges or tuberculate elevations on the escutcheon and the areas, essentially concentric with the lines of growth. The term costae, however, is avoided since in certain of the discussed groups of Trigonia the vertical and subvertical ridges of the flank not only pass into horizontal and subhorizontal ridges, but in some species are U- and V-shaped and inclined at various angles. *Riblets* are the minor solid or tuberculate ridges on the escutcheon and areas which are not essentially concentric with the lines of growth.

Group of *Trigoniae Pseudo-Quadratae*

According to Steinmann (1882, p. 225) the diagnostic characters of the Pseudo-Quadratae are as follows:

The outline of shell, the general nature of ribbing, and the interrelation of areas are as in the Clavellatae. The sum of these features outweighs those connecting the Pseudo-Quadratae with the Quadratae—that is, the resolution of the costellae on the escutcheon and the transition of the concentric sculpture from the ornamented escutcheon across the areas to the anteroventral part of the flank.

In Steinmann's opinion, the forms with equally divided areas should not be included in the Pseudo-Quadratae. The strong development of areal costellae, on the other hand, is regarded as a less important factor, not altogether restricted to the Pseudo-Quadratae and the Quadratae, since it is also known to occur in certain Clavellatae.

Later Kitchin (1903, p. 98) commented that there is no sufficient ground for assuming that the Pseudo-Quadratae actually are transitional forms between the Clavellatae and the Quadratae. He tentatively gave a comparative diagnosis for the Quadratae and the Pseudo-Quadratae:

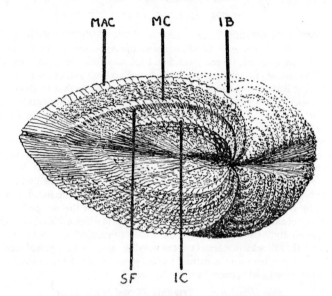

FIGURE 2.—*Carinae and furrows of a Trigonia shell*

IB—Inframarginal Band. MAC—Marginal carina. MC—Median carina. SF—Supramedian furrow. IC—Inner carina.

FOR THE QUADRATAE: Shell short, outline subquadrate, nodose ornamentation on the flanks, areas broad, not steeply inclined to the flanks, indistinctly separated from the latter and from the escutcheon. Escutcheon with nodose or tuberculate ornamentation. Small pits set in rows present near the posterior margin on the inner surface of valves.

Parenthetically, Kitchin's somewhat ambiguous statement that in the Quadratae the areas are not steeply inclined to the flanks evidently is erroneous, because actually in the Quadratae the areas are more steeply inclined than in any other group of Trigoniae.

FOR THE PSEUDO-QUADRATAE: Shell more elongated and tending to approach that of the Clavellatae. Ribs more regularly developed and spaced, and separation of the flank from the areas more effective than in the Quadratae. Pits along the posterior margin in the interior are absent.

Kitchin also pointed out that whereas in the Pseudo-Quadratae the outer area is wider than the inner area, in the Quadratae the inner area is the wider of the two, as observed by Steinmann.

In a later paper, Kitchin (1908, p. 109) lay stress on the relation between the tuberculation of the carinae and the nature of the areal costellae in the Pseudo-Quadratae. Of the well selected examples, in *Trigonia mamillata* Kitchin (1903, p. 100, Pl. 9, figs. 8-9; Pl. 10, figs. 1-3) the stage of three tuberculated carinae ceases very early. It persists farther posteriorly in *T. herzogi* (Goldfuss, *in* Steinmann, 1882, p. 220, Pl. 7, figs. 1-2; Pl. 9, figs. 1-2) and is observed even farther in *T. holubi* Kitchin (1908, p. 103, Pl. 4, figs. 2-2a). Simultaneously with the loss of tuberculation on the median carina, in these species and in *T. transitoria* Steinmann (1881, p. 260, Pl. 13, fig. 3; 1882, p. 221, Pl. 7, figs. 3-4 and Pl. 8, figs. 1-3), the transverse areal costellae become more massive and begin to encroach upon the flank.

Briefly summing up the most diagnostic features of the typical pseudo-quadrate Trigoniae, like *T. transitoria* (Steinmann, 1882, Pl. 8, fig. 3; Burckhardt, 1903, Pl. 14, fig. 1; Weaver, 1931, Pl. 21, fig. 106-108) and *T. herzogi* (Steinmann, 1882, Pl. 7, fig. 1; Kitchin, 1908, Pl. 5, fig. 1): shell more or less elongated, outer area wider than inner area, escutcheon with resolved but areas with solid costellae, areal costellae encroaching upon the flank, and an inframarginal band present, I have to admit that none of the species from Malone, Texas, and from Lower Cretaceous strata of southeastern Arizona can be placed in this group without reservation. Yet, many of the examined species are closer to the Pseudo-Quadratae, as defined, than to the Quadratae, and the gradual modification of certain characteristics clearly determines their relative position nearer the first group. It should be noted in this connection that even within the species of typical Pseudo-Quadratae certain forms have been illustrated that are closer to the transitional species described in this paper than to the types with well expressed characters of the group.

My studies have led me to the conclusion that besides the already discussed diagnostic properties of the shell the following features greatly assist in evaluating relationship of the Trigoniae with mixed characteristics: 1, nature of areal costellae which are entire and untuberculate in the typical Pseudo-Quadratae but are either resolved into tubercles or ornamented with tubercles in the Quadratae; 2, degree and relative position of this resolution; 3, orientation of tubercles on the escutcheon.

It must be admitted at once that the modifications in these two groups of Trigoniae, are of little stratigraphic value *per se*. Thus, for instance, the resolution of areal costellae already indicated in Malone time persists with a gradual advance through the Lowell to the base of the Glen Rose. On the other hand, the species with unresolved costellae also constitute an independent line of development from the Malone to the Lowell. It is noteworthy that the Malone species with entire costellae, subequal areas, and very little infringement of the costellae upon the flank, are much closer to the Upper Aptian forms of Arizona than to the admittedly Neocomian-Lower Aptian Trigoniae of the so-called southern realm, in which the pseudo-quadrate characteristics, like the unequal areas, encroachment of the costellae, weakly developed median carina, inframarginal band, and orientation of tubercles along the lines of growth on the escutcheon are in the full display. The more elongate Lower Cretaceous Pseudo-Quadratae of South America seem to be closer to the Clavellatae. In *Trigonia transitoria* Steinmann (1882, Pl. 8, fig. 3), for instance, the outer area is considerably wider than the inner area; the median carina is barely perceptible even in its anterior part and bears no prominent tubercles; the infringement of the posterior costellae and inframarginal band are well expressed; the tubercles on the escutcheon are not set in horizontal rows (Steinmann, 1882, Pl. 8, fig. 2). Assuming the correct interpretation of their age, it may be surmised that Trigoniae of the typical *transitoria* character are altogether restricted to the Lower Cretaceous of the South Andean and Africano-Indian province.

Taken in its entirety, however, the species of pseudo-quadrate Trigoniae clearly indicate a gradual trend in the development culminating in the Quadratae. As has been mentioned before, even in the assemblages originally referred to the Pseudo-Quadratae there are among the typical forms certain representatives which are closer to the Texas and Arizona species. Thus, the smaller of the two specimens described and illustrated by Steinmann (1882, Pl. 8, fig. 1) as *Trigonia transitoria* exhibits no inframarginal band but has *paired costellae* in the anterior part of both areas, and the shell itself is less elongate than the other form (Steinmann, 1882, Plate 8, figure 3) which has all the characters of Steinmann's diagnosis and should be taken as the lectotype of his species. Further, in the very elongate shell of *Trigonia herzogi* (Goldfuss) the median carina is fairly strong in the anterior part and dwindles only posteriorly (Steinmann, 1882, Pl. 7, fig. 1; Kitchin, 1908, Pl. 5, fig. 1). *Trigonia transitoria* var. *quintucoensis* Weaver (1931, Pl. 23, figs. 119–121) approximates *Trigonia vyschetzkii* Cragin (1905, Pl. 8, fig. 2) and likewise has the tubercles on the escutcheon aligned in transverse rows in its posterior part.

Among the species with unresolved areal costellae *T. vyschetzkii* Cragin (1905, Pl. 8, fig. 2, only) from Malone, geologically the oldest known pseudo-quadrate Trigonia of North America, has neither a smooth inframarginal band nor a well-developed encroachment of the costellae upon the posterior part of the flank. Its outer area, however, is somewhat wider than the inner area. The median carina, well developed and even tuberculate in the anterior part, is barely indicated posteriorly where it is overshadowed by the supramedian furrow. In the anterior part of the escutcheon the tubercles are set in transverse rows. In *Trigonia guildi*, sp. nov. (Pl. 12, figs. 1–2) from the Upper Aptian of southeastern Arizona the shell is less elongate than in *T. vyschetzkii*, the areas are subequal, the median carina is well developed and provided with tubercles for two anterior thirds of its length, and the areal costellae are paired. As in the latter species, there is no inframarginal band and the tubercles are set in transverse rows in the posterior third of the escutcheon. In both species the areal costellae cross the marginal carina in its posterior part.

No pseudo-quadrate Trigoniae with finely resolved areal costellae have ever been described from the Lower Cretaceous of South America, South Africa, and British India. However, the Jurassic *Trigonia mamillata* Kitchin (1903, p. 100, Pl. 9, figs. 8–8a) from the Oomia (Umia) strata of British India has coarse tubercles on the costellae of the outer area and to a lesser degree even in the inner area. Except for this areal ornamentation and a more strongly tuberculated median carina, this interesting Trigonia has all essential characters of the *transitoria* group. The peculiar reduction or even absence of the marginal carina remarkably accentuates the inframarginal band in the anterior part of the shell.

Among American Trigoniae with resolved costellae, *Trigonia dumblei*: nom. nov., (Pl. 12, fig. 4) from Malone is a close relative of *Trigonia resoluta*, sp. nov. (Pl. 12, fig. 3), from the Lowell formation of Arizona. Both species have the costellae resolved only in the anterior part of the outer area. In the Arizona species the areal costellae are less crowded and their resolution is slightly more advanced posteriorly.

Trigonia saavedra, sp. nov. (Pl. 13, fig. 5), from the Upper Aptian of Arizona, of more subquadrate outline, has resolved costellae in the anterior part of both areas, the areas more steeply inclined, and the outer area narrower than the inner area, more as in the Quadratae. As in the Quadratae, the ribs in the umbonal portion of the flank are strongly bent in a V-shaped and U-shaped manner, contrary to their straight course in the Pseudo Quadratae; Steinmann (1882, p. 220) considered such difference of diagnostic value in discriminating between the two groups.

Trigonia mearnsi, sp. nov. (Pl. 12, figs. 5–6; Pl. 13, figs. 1–4), from the Upper Aptian of Arizona and Texas, with its subquadrate outline, flatter flanks, narrow escutcheon, rapidly broadening areas, and areal costellae resolved in the two anterior thirds of both areas, is closer to the Quadratae than any other species discussed. Nevertheless, the total absence of large nodes or of deviating and zigzagging costellae in its areal ornamentation, the lack of alternating pits and elevations along the posterior margin on the inside of the valves, and especially the lack of uninterrupted concentric ornamentation across the entire valve from the dorsal to the ventral margin, place this species closer to the general trend of development of the Pseudo-Quadratae. It is also significant that in *T. mearnsi* the tubercles on the escutcheon are not arranged in transverse rows but follow the lines of growth exactly as in the earlier Pseudo-Quadratae. (*Compare* Steinmann, 1882, Pl. 8, fig. 2.)

T. vyschetzkii and *T. guildi* are closer to the Pseudo-Quadratae than are *T. dumblei* and *T. resoluta* with resolved costellae in the outer area. *T. saavedra* and *T. mearnsi* with the resolution of areal costellae advanced in both areas are much closer to the Quadratae, although they retain certain characteristics not in evidence in the typical members of the last group.

Regarding the relation between the development and tuberculation of the carinae, on the one hand, and the accretion of the costellae, on the other, as emphasized by Kitchin, it must be stated that none of the American species referred to the Pseudo-Quadratae in this paper has the condition of *Trigonia transitoria* Steinmann in which the untuberculated median carina is barely indicated, whereas the solid costellae are strongly sculptured in the posterior part of the areas. Nevertheless, it appears that in *T. vyschetzkii* the median carina is less developed and tuberculated than in any other discussed American species, whereas in *T. mearnsi* it is tuberculated almost to the posterior margin, in agreement with the reduction of the areal costellae which in the posterior region of the last named species are much less prominent and often completely merge with the lines of growth.

Status of Trigonia vyschetzkii Cragin

There are two sets of specimens in the U. S. National Museum which were collected by Cragin and his associates at Malone, Texas, and referred by Cragin to his species *Trigonia vyschetzkii*: (1) an older set, on which the original diagnosis of the species is supposed to have been established (Cragin, 1893, p. 215) and of which not a single specimen was illustrated, and (2) a later described and illustrated assemblage widely known in the literature as *Trigonia vyschetzkii* Cragin (1905, p. 56, Pl. 8, figs. 1–2; Pl. 9, figs. 1–3).

The older set is provided with two printed labels:

> 215
> *Geological Survey of Texas*
> Trigonia vyschetzkii Cragin
> Part of type material
> *Locality* El Paso county Texas
> Collector R. Vyschetzkii

and

> *Geological Survey of Texas*
> Trigonia vyschetzkii Cragin

This older set consists of four very weathered specimens with green rectangular markers, with printed "506" pasted on each of them. Besides, each specimen has a brown rectangular marker with "483" in ink, and two of the examples also bear white rectangular labels with "215" also in ink. There are no individual identification markings on any of these specimens, and therefore the specimens of this set cannot be referred to without individual description.

None of these forms, partially or supposedly representing *T. vyschetzkii* of Cragin's earlier description, can be considered as the type of that species. Every one is in such fragmentary or weathered condition that to none could be applied the definition of Cragin's original diagnosis: "Shell large ... subovate ... describing ventrally a sweeping curve ... area narrow, its ornamentation consisting of a double avenue of narrow, elevated, obliquely transverse, and more or less beaded and tubercularly terminated costellae ...," etc. The ventral portion of the shell is preserved in none of the specimens; the areal costellae are not distinguishable; there are only barely perceptible traces of costellae in some of them. Beyond doubt, the original description was made from other specimens, most probably from the set described and illustrated later by Cragin in the well-known Malone Bulletin (Cragin, 1905). The only large example of this older collection is represented by a portion of the flank with four or five ribs preserved. These ribs bear large tubercles which seem to be set closer than in the ribs of Cragin's *T. vyschetzkii* illustrated in Figure 1 of Plate 8 of the Malone Bulletin. This, however, may be a result of weathering which always tends to make tubercles of a Trigonia appear flatter, lower, and more closely set.

Since Cragin's original description of *T. vyschetzkii* (Cragin, 1893) was quite evidently not made from any of the four specimens of the older set, or from all of them, and because none of them can be identified with any degree of assurance with the material described and illustrated by Cragin as *Trigonia vyschetzkii* in his paper of 1905, these specimens are not to be considered in deciding upon what should be the type of Cragin's species. In my opinion the original diagnosis of *T. vyschetzkii* was made by Cragin from the specimens later described in the Malone Bulletin, presumably from the forms illustrated in Figure 2 of Plate 8 and in Figure 1 of Plate 9.

I also have had the opportunity to examine all "cotypes" of *T. vyschetzkii* in the Museum of Economic Bureau of Texas at Austin. These specimens are marked "506-583." I have observed that these better preserved specimens have: (1) costellae that cross the marginal carina but do not appreciably encroach upon the flank, (2) ribs that reach the anterior border of the shell, and (3) tuberculate costellae on the escutcheon that are set normal to the cardinal margin.

The "later" set, that is, the set of illustrated description of 1905, is preserved in the National Museum in two boxes. Each box is provided, as is each of the specimens, with a green diamond marker. The printed label in the box with three specimens illustrated by Cragin (1905, Pl. 8, figs. 1 and 2, and Pl. 9 fig. 2) reads:

> *U. S. N. M.*
> 28967
> Trigonia vyschetzkii Cragin
> Jurassic
> Malone
> Texas
> *Identified by* F. W. Cragin
> Bull. U. S. G. S. No. 266. p. 56, pl. 8, f. 1. 2; pl. 9, f. 2–3

The other box, with the specimen figured by Cragin in Pl. 9, fig. 1, has the following label:

```
U. S. N. M.
   29020
           Trigonia vyschetzkii Cragin
                    Jurassic
                 1½ miles E. of Malone
                                  Texas
     Identified by F. W. Cragin
     Bull. U. S. G. S. No. 266, p. 56, pl. 9, fig. 1
```

These specimens can further be identified by the markers pasted on them and numbers written in ink directly on the specimens.

Thus, the specimen illustrated by Cragin (1905) in Figure 1 of Plate 8 and in Figure 3 of Plate 9 has two green diamond markers, a white rectangular marker with "1" written in ink, and "28967" in ink written directly on the specimen. This specimen will be referred to in the present description as No. 1.

The form illustrated in Figure 2 of Plate 8 of the same publication has a green diamond, a number "28967" in ink on the specimen (barely legible), and a square white paper with "3" on it. This is No. 2 of the present description.

The specimen illustrated by Cragin in Figure 1 of Plate 9 has a green diamond, a green circular marker with "1807" in ink (the last figure rather indistinct), a white rectangular marker with "2" in ink, and "29020" written directly on the specimen. This form is to be referred to as No. 3 of Cragin's second set of Trigoniae from Malone.

The specimen shown in Figure 2 of Plate 9 of the same publication has a green diamond, and "28967" in ink on the specimen (the second and the fourth figures very indistinct), and "4" written in ink on a white square marker. This specimen is designated as No. 4 in the further discussion.

Not all these forms are conspecific, as may be inferred from the following review. Part of this material is appreciably close to *Trigonia guildi*, sp. nov., from the Lowell formation of Arizona.

Trigonia sp.

1905. *Trigonia vyschetzkii* CRAGIN (part), Paleontology of the Malone Jurassic formation of Texas, U. S. Geol. Survey, Bull. 266, p. 56, Pl. 8, fig. 1, and Pl. 9, fig. 3, only.

Specimen No. 1 of Cragin's illustrated collection. Ten coarse ribs with large tubercles or nodes present on flank, and a part of an earlier rib in umbonal region. Ribs on beak not preserved. Dorsal part of left valve (only internal mold of right valve is left) is broken off for three posterior quarters just below marginal carina, as it is well seen in Cragin's figure. Marginal carina was undoubtedly placed more dorsally than in specimen No. 2.

The areas and the escutcheon are preserved only in the anterior part of the shell, but their ornamentation cannot be satisfactorily interpreted because of eroded condition. They are horizontal in this part of the specimen. Judging from what is preserved of the areas, it may be safely concluded that they were much less inclined on the entire shell than either in the specimen No. 2, or in *T. guildi*, which would agree with the postulated (more dorsal) position of the marginal carina. Only a faint line representing the median carina is visible; nevertheless, its presence makes it clear that both areas were not of equal width in their anterior portion. The inner carina bears tubercles, five of which are observable. These tubercles are of the same nature as in the specimen No. 2 and in *T. guildi*, that is, rather transverse across the carina than round. It is quite apparent that the costellae on the escutcheon are resolved into tubercles set in rows normal to the cardinal margin. The number of these tubercles, not less than three, possibly four in a row, and their nature, also appear to be the same as in the Arizona species.

This specimen cannot be conspecific with specimen No. 2 because of the high position of the marginal carina, the nearly horizontal areas and escutcheon, the outer area being wider than the inner area already in the anterior part of the shell, and more inflated flanks. I have four specimens from Malone with the same characters of the areas and of the marginal carina. None of them, however, has the ornamentation of the areas and escutcheon sufficiently preserved to allow a specific description.

Trigonia vyschetzkii Cragin, emend.

1905. *Trigonia vyschetzkii* CRAGIN (part), Paleontology of the Malone Jurassic formation of Texas, U. S. Geol. Survey, Bull. 266, p. 56, Pl. 8, fig. 2, only. Holotype by designation.

Specimen No. 2 of Cragin's illustrated collection. This form is the nearest approach to *T. guildi* of all Trigoniae described by Cragin as *T. vyschetzkii*. Cragin's description of this species was quite apparently made essentially from this specimen. Its flank is considerably weathered, with the part above the antero-ventro-posterior margin broken off, but the areas, carinae, and the anterior part of the escutcheon are satisfactorily preserved.

Compared to *T. guildi*, the shell of this species is slightly longer and posteriorly somewhat flatter. Its ventral margin forms a longer and gentler curve. The ribs on the flank are less numerous, about ten against eleven or twelve of the Arizona species. These ribs bear larger nodes, although the number of nodes or tubercles appears to be the same on the corresponding ribs of both species. In *T. vyschetzkii* the ribs probably had a somewhat less pronounced forward sweep in the anteroventral region which is destroyed in the specimen.

The position of all three carinae on the flank and the relative width of the inner and outer areas in their anterior part are approximately the same in both species, but the outer area of the Texas form is less inclined and becomes appreciably wider in its posterior part. As in *T. guildi*, the marginal carina of *T. vyschetzkii* is ornamented with tubercles which gradually become less distinct and eventually cease posteriorly. The tubercles of the median carina are in evidence only in the anterior part of the latter form but the entire carina dwindles posteriorly and is superseded by the rapidly developing supramedian furrow, as in *T. transitoria* Steinmann. In the Arizona species, on the other hand, the corresponding tubercles extend farther posteriorly in this way marking the median carina limited above by the supramedian furrow.

One of the distinguishing features of *T. vyschetzkii* is the relative position of the costellae of the inner and outer areas. As in *T. transitoria*, in this species the costellae of the inner area pass into the costellae of the outer area with a very insignificant change in direction, and therefore the angle between the two is wide and barely perceptible. It may be said that the costellae of both areas are uniformly directed posteriorly. The costellae of *T. guildi*, on the other hand, are oriented less posteriorly on the outer area, and consequently form a distinct angle with the costellae of the inner area. The latter are directed as much posteriorly as the costellae of Cragin's species.

The encroachment of the areal costellae upon the flank in the posterior part of the shell of *T. vyschetzkii* described by Cragin as "....folds [costellae]...in some cases even passing a little beyond it [marginal carina]....," with a conclusion that "...some—usually slight—tendency which the areal folds [costellae] manifest to infringe on the general surface [flank] below the limiting angle [marginal carina], on the posterior region...," brings his species in agreement with the Quadratae (Cragin, 1905, p. 56, 57), is not seen in any types of Cragin's primary and secondary sets. This encroachment is present to some extent in the holotype of *T. guildi*. Paratype No. 91413 (not illustrated) shows the areal costellae encroached upon the flank as far as in *T. transitoria* Steinmann (1882, Pl. 8, fig. 3) and *T. herzogi* (*in* Steinmann, 1882, Pl. 7, fig. 1); as in these two species, the costellae of the paratype pass ventrally on the flank into elongated nodes, which, although a continuation of the areal costellae, are considerably thicker and merge into shorter tubercles—the integral parts of the ribs.

The middle and posterior parts of the escutcheon in the holotype of Cragin's species are badly weathered and do not show the ornamentation. In the preserved anterior part of the escutcheon the tubercles are set in transverse rows.

T. vyschetzkii differs from typical Pseudo-Quadratae, as exemplified by *T. transitoria* Steinmann, in having tubercles on the escutcheon arranged in rows normal to the cardinal margin, instead of being aligned along the lines of growth, and in having a partly tuberculated median carina. The smooth inframarginal band, so characteristical of the Pseudo-Quadratae (Steinmann, 1882, Pl. 8, fig. 3; Weaver, 1931, Pl. 24, figs. 126, 128, and 129), is also lacking in Cragin's species.

The characteristics which *T. vyschetzkii* has in common with the Pseudo-Quadratae and by which it differs from the Upper Aptian species of Arizona are: outer area somewhat broader than inner area, median carina early replaced by supramedian furrow, costellae of both areas forming continuous lines, and ribs ornamented with larger nodes and having a less pronounced forward sweep in anterior region.

Trigonia maloneana Stoyanow, nom. nov.

1905. *Trigonia vyschetzkii* CRAGIN (part), Paleontology of the Malone Jurassic formation of Texas, U. S. Geol. Survey, Bull. 266, p. 56, Pl. 9, fig. 1, only. Holotype by designation.

Specimen No. 3 of Cragin's illustrated collection of "*T. vyschetzkii.*" This Trigonia is distinguished from all other representatives of the Pseudo-Quadratae by very concave, nearly crescentic, dorsal margin with rapidly rising umbo and posterodorsal extremity. This is not satisfactorily conveyed by Cragin's photographic illustration in which the umbonal part is foreshortened and the posterior margin contracted. The interrelation of the areas, on the other hand, is closer to that in the typical species of the Pseudo-Quadratae, like *Trigonia transitoria*, than in any other Trigonia of Texas and Arizona. In the holotype the outer area is 20.5 mm. wide at its posterior termination, whereas the width of the inner area is there only 9 mm.

I have a good topotype of this species with a better preserved posterior part of the shell. The present description, however, is based entirely on the right valve (the only one preserved) of the holotype. The areas form a more obtuse angle with the escutsheon than in *T. vyschetzkii* and *T. guildi*; they also do not make a continuous slope with the flank as in those two species. On the contrary, the areas tend to be concave posteriorly in a striking contrast to the inflated flank.

In the foremost part of the escutcheon there are a few fine elongate tubercles which seem to be aligned parallel with the lines of growth. However, very early the tubercles become arranged in rows which are normal, in places nearly normal, to the cardinal margin. Seven of such rows are preserved. Cragin's illustration of this form does not give an adequate and correct idea of the true nature of these tubercles. They are not angular or wedge-shaped, as is seen in the picture. Instead, the escutcheonal tubercles are round with tendency toward a posterior elongation, as in *T. vyschetzkii* and *T. guildi*. The first three rows have four tubercles each; the fourth and the fifth rows are composed of five and probably (the fifth row) of six tubercles. Their number in the last row cannot be definitely established. Incidentally, No. 3 is the specimen on which Kitchin (1926, p. 462) commented in his discussion of the ornamented escutcheon of *T. vyschetzkii*.

The true ornamentation of the inner carina also is not adequately represented in Cragin's illustration. A few anterior "tubercles" are mere elongated elevations directly passing into the corresponding costellae of the inner area. Very early these elevations become so oriented on the carina as to form broad V-s with the rows of tubercles of the escutcheon, and broad inverted V-s with the costellae of the inner area.

The course of the costellae as they pass from the inner area onto the outer area is very uniform through the observable length of the shell. There is a weakening or slight interruption above the median carina (not as much furrowed as seen in Cragin's figure) with a barely perceptible angulation between the inner and outer costellae, and a swelling of the costellae on the median carina. These swellings, or elongate tubercles, do not interfere with the general trend of the costellae.

On the marginal carina the outer costellae terminate in elongate tubercles which are less oblong in its anterior than in the posterior part.

The ribs are of the same type as in *T. vyschetzkii*, but probably not as sparsely set in the anterior region of the shell.

The strongly inflated anterior part of the shell with a rather heart-shaped outline suggestive of scabroid Trigoniae, and the V-shaped arrangement of tubercles and costellae on the escutcheon and the areas, suggestive of that in the Quadratae, separate this species from *T. vyschetzkii* and probably from the Pseudo-Quadratae in general.

Trigonia maloneana Stoyanow, var.

1905. *Trigonia vyschetzkii* CRAGIN (part), Paleontology of the Malone Jurassic formation of Texas, U. S. Geol. Survey, Bull. 266, p. 56, Pl. 9, fig. 2, only. Holotype by designation.

Specimen No. 4 of Cragin's illustrated collection. This specimen is very fragmentary. In common with *T. maloneana* it has areal costellae that form inverted V-s as they pass from the inner carina to the inner area. There are, however, fine tubercles in the posterior part of the inner area which suggest a resolution of costellae in that part. Ornamentation of the escutcheon consists of oblique ridges which, closer to the cardinal margin, give way to tubercles set along the lines of growth. Three smaller specimens of Cragin's primary assemblage seem to have a similar ornamentation on the escutcheon but they are too eroded to allow a definite interpretation.

Trigonia guildi Stoyanow, sp. nov.

(Plate 12, figures 1-2)

Shell subovate in general outline, gradually inflating toward the middle from the anterior and posterior margins, with the greatest thickness slightly anterior to the middle of the flank, appreciably compressed posteriorly. Anterior margin, little curved in its dorsal part, passes into gently curved ventral margin, with which it is continuous, and which in its turn passes without interruption into posterior margin. The latter is somewhat truncated dorsally. Escutcheon slightly inclined outward, ornamented with tubercles which anteriorly are very elongated and oriented along the lines of growth, but posteriorly become wider, more rounded, and are arranged in rows sub-normal to the cardinal margin composed of three tubercles each. Areas more inclined than the escutcheon. All three carinae bear round tubercles in their anterior part. In the posterior part of the shell, however, the tuberculation ceases, and the carinae are crossed by the costellae only. Supramedian furrow well developed, especially posteriorly. Inframarginal band absent. Areal costellae directed slightly posteriorly in the inner area, and as much anteriorly in the outer area, in this way forming a very wide angle bisected by the median carina. These costellae are double, that is, there is a frequent repetition of two closely set costellae, one beginning from the anterior and the other from the posterior sides of the carinal tubercles. In places there are intervening costellae which cross the median carina between the tubercles; sometimes such secondary costellae occur also within the paired costellae. Such arrangement is more regularly developed in the outer than in the inner area. In the posterior part of the shell the costellae cross the marginal carina and encroach upon the flank forming elongated nodes. Ribs are little directed anteriorly except for forward sweep in the anteroventral part of the shell. Simultaneously with this sweep the round tubercles adorning the ribs become transversally elongated nodes.

Compared to *T. vyschetzkii*, as defined in this paper, *T. guildi* has a more rapidly curving ventral margin. The relative width of the areas is nearly the same in the anterior part of both species, but in *T. vyschetzkii* the outer area is appreciably wider in the posterior part of the shell. In the latter species the areal costellae are single and do not form an angle as they pass from the inner to the outer area. I was inclined to believe that the single areal costellae of the Texan form might have resulted from somewhat weathered condition, but recently collected topotypes show this character as constant for Cragin's species. Further difference between the two discussed species is seen in a weaker development of the median carina in *T. vyschetzkii* which, ornamented with weak tubercles only anteriorly, ceases altogether in the posterior part of the shell; in a lesser number of more distantly spaced ribs which apparently lack a well pronounced forward sweep in the anteroventral region; and in relatively larger tubercles on the ribs and longer nodes on the flank.

T. guildi disagrees with the Pseudo-Quadratae in having the tubercles on the escutcheon set in more or less horizontal rows posteriorly, and not obliquely on the lines of growth, as in *T. transitoria* (Steinmann, 1882, Pl. 8, fig. 2) and in *T. neuquensis* Burckhardt (*in* Weaver, 1931, Pl. 22, fig. 114), in lacking the inframarginal band, in having the areas of equal or subequal width, and in the limited encroachment of the areal costellae upon the flank. *T. guildi* agrees with *T. transitoria* in having paired costellae (*see* part of Steinmann's cotypes, 1882, Pl. 8, fig. 1, in the anterior region only) although in the Arizona species such costellae are present farther posteriorly. The median carina is well developed in the described species and bears tubercles as far as its posterior third, whereas in *T. transitoria* this carina is barely perceptible, without a distinct tuberculation, or is absent altogether. However, in another familiar example of the Pseudo-Quadratae, *T. herzogi* (Goldfuss, *in* Steinmann, 1882, Pl. 7, fig. 1), this carina is tuberculated for two thirds of its length.

TYPE: Holotype: No. 91205. Paratype: No. 91413 (not illustrated).

OCCURRENCE: This species is abundantly represented in lower part of Cholla member of Lowell formation, Ninety One Hills. Holotype and paratype collected from division 6a.

Trigonia resoluta Stoyanow, sp. nov.

(Plate 12, figure 3)

This species differs from *Trigonia guildi* in having the areal costellae resolved into tubercles in the anterior part of the outer area. The ribs are set closer and the tubercles ornamenting them are smaller than in *T. guildi*.

The resolution of areal costellae is a common feature in the Quadratae. Such costellae are present, for instance, in *Trigonia hondaana* Lea, as illustrated by Dietrich (No date, 1938?, Pl. 18, figs. 1a, 6, etc.). In the species of the Quadratae with resolved costellae this resolution is more advanced posteriorly, and in the mentioned South American species the entire areal ornamentation is of a different pattern because of the riblets set at an angle to the costellae in the anterior part of the outer area.

Trigonia resoluta is much closer to a form from Malone which originally was described by Cragin as one of the two cotypes of his species *Trigonia taffi*, but later removed by him to *T. vyschetzkii* (see Pl. 12, fig. 4). The name *Trigonia taffi* was restricted by Cragin to the remaining specimen, the holotype. The name of the latter species, which has never been illustrated, is widely used in the geological literature of Texas. A species from the Glen Rose of Bluff Mesa and Mayfield Canyon in the Quitman Mountains, abundantly represented in paleontological collections of Texas and commonly labeled as "Trigonia taffi," is not conspecific with either of the two of Cragin's types, and is described in this paper as *Trigonia mearnsi*, sp. nov.

TYPE: Holotype: No. 91418.

OCCURRENCE: Holotype was collected from Cholla member, division 6a, of Lowell formation, Ninety One Hills.

Status of *Trigonia taffi* Cragin

In the U. S. National Museum two specimens of Trigoniae are preserved, a large one and a small one, which were originally described by Cragin as *Trigonia taffi*. The large specimen has a label:

Trigonia taffii Cragin
        Part of type material
El Paso county.   Texas
W. H. Streeruvitz

The markers on this Trigonia bear the numbers: 214, 994, and 509 (green, apparently a museum label). This specimen will be referred to as No. 509.

The label accompanying the smaller specimen reads:

(One of the type-specimens)
Trigonia taffi Cragin
1 mi. N. E. of Malone
El Paso Co., Tex.
W. H. Steeruvitz

There are the following markers on this Trigonia:

214	483	511
White paper	Brown paper	Green paper
Brown ink	Black ink	Print

This specimen is discussed as No. 511.

According to the label and to Cragin's original description of "*Trigonia taffi*" (Cragin, 1893, p. 214) the smaller specimen No. 511 was found 1 mile northeast of Malone. Referring to the finding place of *Trigonia vyschetzkii*, Cragin (1893, p. 215) says: "Collected by Messrs. von Streeruwitz and Wyschetzki, one mile northeast of Malone. There abundant and associated with *Trigonia taffi* [sic!], *Venus malonensis*, and other bivalves named under the latter species."

On the other hand, the large specimen No. 509 was collected on Bluff Mesa in the Quitman Mountains (Cragin, 1893, p. 214).

In a footnote on page 10 of the Malone paper Cragin (1905) says:

"The type material of *Trigonia taffi* was described (loc. cit., p. 214) as having come in part from Bluff Mesa and in part from the locality of *Trigonia vyschetzkii*, east of Malone. It is now, however, practically certain that the nearly complete valve from Bluff Mesa alone represented *Trigonia taffi*, and that all of the fragmental material which was supposed to belong to this species, and which was from the Malone locality, belonged—as part of it which I recently had opportunity of reexamining certainly does—to *T. vyschetzkii*."

This is an important statement from which it follows:

1. The type of *Trigonia taffi* Cragin is the large specimen No. 509 from Bluff Mesa, a holotype.

2. The smaller specimen No. 511 comes from Malone, and was collected together with *T. vyschetzkii*, in fact, was identified by Cragin with that species. A careful examination of specimen No. 511 shows that the species it represents is much closer to *T. resoluta* than to *T. vyschetzkii*.

Very unfortunately, Cragin, maintaining in his original description (1893) that both forms represented one species, did not sufficiently discriminate between them. It is quite evident, nevertheless, that only the larger specimen furnished the measurements given in the "Contribution to Invertebrate Paleontology of Texas Cretaceous" (1893, p. 214 and 215), but the description was undoudtedly made from the smaller and, in my opinion, better preserved specimen. The type No. 511 is discussed and illustrated in this paper as *Trigonia dumblei*, nom. nov., a holotype, (Pl. 12, fig. 4). Regarding the specimen No. 509, the holotype of *Trigonia taffi*, Cragin (1905, footnote, p. 10) mentioned that it is represented by a "nearly complete valve." Actually, in this *large* specimen from Bluff Mesa only the flank and the middle portion of the outer area are observable, notwithstanding the fact that the entire valve indeed is preserved. It is definitely manifest that the diagnostic characteristics of "*Trigonia taffi*," as outlined by Cragin, like the resolution of the areal costellae and nature of the ribs, cannot pertain to the holotype No. 509 in which the visible costellae are entire, and the ribs in the anterior part of the valve, in a marked contrast to *T. dumblei*, have a forward sweep very high on the flank, almost immediately below the marginal carina, a condition which persists to the midlength of the shell. In addition, in *T. taffi* the posterior ribs zigzag and the tubercles are longitudinally elongate in the ventral part of the valve, a feature not mentioned by Cragin. Incidentally, in the holotype of *T. dumblei*, from which the description was made, the ventral portion of the shell is not preserved.

In his diagnosis, which pertains to *T. dumblei* and not to *T. taffi*, Cragin pointed out that:

(1) The areal costellae are resolved into tubercles only in the anterior part of the area (the outer area only)—"near the beaks."

(2) The areal costellae are separated by interspaces about three times as broad as the costellae.

(3) The flanks are ornamented with rather small tubercles so set as to constitute interspaced tubercular ribs, the tubercles being uniform in size and round.

T. resoluta is close to *T. dumblei*. The relation between the two species is as follows:

(1) The resolution of costellae in the anterior part of the outer area is identical in position and nature. In the Arizona species it is slightly more advanced posteriorly.

(2) In both species the areal costellae are paired. Their number is less in *T. resoluta*, in which in the part of the outer area immediately following the anterior portion with resolved costellae there are 14 paired costellae within 1 inch. There are 20 such costellae within the same part of the shell in *T. dumblei*. Consequently, the interspaces between the costellae are wider in *T. resoluta*.

(3) The ornamentation on the flank is similar in both species. The tubercles on the midflank of *T. dumblei* appear more flat-topped and lower because this part of the holotype suffered more from weathering than the anterior and posterior parts. In both species ribs are subconcentric on the umbones and gently curved on the flank. A rib on the midflank, approximately the eleventh in *T. resoluta* and the fourteenth in *T. dumblei*, is rather outstanding because it is directed backward straight from the marginal carina and separates the anterior ribs from the posterior set of ribs. The latter are oriented more backward in *T. dumblei*. The escutcheon is not preserved satisfactorily in either of the holotypes. The inner area is not adequately exposed in *T. dumblei*, but appears to be relatively narrower than the outer area. In the Arizona species the areas are subequal.

Essential difference between the two species is in more numerous and crowded areal costellae, and a stronger backward inclination of the ribs in *T. dumblei*. In *T. resoluta* the costellae are more prominent, but this may be due to a better preservation of the specimen.

Trigonia saavedra Stoyanow, sp. nov.
(Plate 13, figure 5)

This species is represented by abundant but imperfect small specimens of subquadrate general outline, more or less depressed in the dorsal, posterior, and ventral regions, and rapidly inflating toward the middle of the shell with the greatest thickness anterior to the midflank at the marginal carina.

The areas are equally and rather steeply inclined. All three carinae bear tubercles except in the posterior part of the shell where the areal costellae, directed posteroventrally and forming a very wide angle as they pass from the inner to the outer area, cross the median carina without producing round tubercles. The median carina is but little concentric with the marginal carina posteriorly. Anteriorly it tends to approach the marginal carina rather closely, often merging with it before the anterior fourth of the shell is reached. Consequently, the outer area appears much narrower than the inner area.

The areal costellae are paired in the posterior part of the shell and do not encroach upon the flank. In the anterior half of both areas the costellae are resolved into fine small tubercles set in rows which preserve the course of the costellae.

The escutcheon is narrow and ornamented with tubercles, not sufficiently preserved in collected specimens for detailed observations.

The ribs are strongly V-shaped in the umbonal part of the flank, and somewhat U-shaped farther ventrally. In the middle and posterior parts of the flank they are markedly directed backward.

This Trigonia is interesting because it has certain characteristics that bring it closer to the Quadratae. However, all areal costellae, whether resolved or entire, are oriented strictly posteroventrally from the inner carina, without interference in this arrangement in all studied specimens.

TYPE: Holotype: No. 91333. Paratypes: Nos. 91320, 91321, 91334, (not illustrated).

OCCURRENCE: Abundant in Saavedra member, division 7n, of Lowell formation, Ninety One Hills.

Trigonia mearnsi Stoyanow, sp. nov.
(Plate 12, figures 5-6; Plate 13, figures 1-4)

Shell large in the full grown stage, subquadrate in outline, moderately inflated, with the greatest thickness in the umbonal third at the marginal carina slightly above the midheight, considerably compressed dorsally. In the umbonal part of the shell the flanks and areas have a common ornamentation which, however, does not extend upon the escutcheon. It consists there of seven initial ribs, of which the first two are broadly V-shaped, and the rest undulating across the flank and the areas to form flat-bottomed W-s between the anterior margin of the shell and the inner carina. Escutcheon very narrow, ornamented with tubercles aligned along the lines of growth and not set in rows normal to the cardinal margin. These tubercles are very small and elongated anteriorly but appreciably increase in size and become more rounded in the posterior part of the escutcheon. Areas subequal, inner area slightly broader, especially anteriorly. All three carinae strongly developed, inner carina with large nodes oriented vertically along the lines of growth, median and marginal carinae adorned with more rounded tubercles. Areal costellae resolved into fine small tubercles in the anterior half, sometimes in the anterior two thirds, of both areas. Posteriorly costellae are entire and closely set, they are slightly wavy, cross the median carina to form a very broad undefined angle, in places with nodular swellings on the carina, and do not encroach upon the flank. Ribs on the flank, except in extreme umbonal region, are directed anteroventrally with a forward sweep in the anterior part of the shell and sometimes with an upward swing closer to the anterior margin. They are formed of closely set round tubercles in the upper part of the flank nearer the marginal carina, but ventrally and anteriorly their course is indicated by the position of more isolated, very large, elongated nodes which become proportionately narrower and smaller toward the anterior margin.

I have examined the inside of the valves in a number of specimens of this species. The small quadrate pits alternating with elevations between the pallial line and the posterior margin, considered characteristic for the Quadratae (Lycett, 1879, p. 100; Steinmann, 1882, p. 221, Pl. 9, fig. 1), are absent in the Arizona species, likewise the depressions and elevations in the posterior part of the ventral margin illustrated by Lycett (1879, Pl. 24, fig. 1a) in his standard example, have not been observed.

Trigonia mearnsi has many characteristics in common with the well-known species of the Quadratae. The general shape of the shell, nature of initial ribs, and inclination of the areas, are very similar to those of *T. daedalea* Parkinson and *T. nodosa* Sowerby (*in* Lycett, 1879, Pl. 22, figs. 7-8; Pl. 23, figs. 1-3; Pl. 24, figs. 1-2). On the other hand, the typical representatives of the Quadratae are separated from *T. mearnsi* by the weak development of the carinae, irregular and variable pattern of areal ornamentation, and concentric sculpture often composed of uninterrupted units passing from the dorsal to the ventral margin across the entire valve. The alignment of the tubercles on the escutcheon along the lines of growth and not in transverse rows normal to the cardinal margin strongly connects the Arizona species with the Pseudo-Quadratae.

Specimens of *T. mearnsi* are abundant in the Perilla member of the Lowell formation in the Bisbee area and in Guadalupe Canyon near Sycamore Creek, 5 miles north of Halls Ranch, in southeasternmost corner of Arizona. In the latter locality they were collected in 1892 by E. A. Mearns of International Boundary Commission, and are now deposited in the U. S. National Museum labeled as *Trigonia taffi* Cragin.[22] The casts of some of Mearns' specimens were kindly sent to me by Dr. Reeside. They have the following label:

Trigonia taffi Cragin
  Loc. 1691
  Guadalupe Canyon near Sycamore Creek
  5 miles north of Halls Ranch, S. E.
  Arizona
              Trinity group

Trigonia mearnsi is also abundantly represented in the Quitman Mountains, Texas. In Mayfield Canyon well preserved specimens are found with *Trinitoceras* below the *Orbitolina* zone.

TYPE: HOLOTYPE: No. GU46. PARATYPES: Nos. GU42, GU45, GU47, GU50.

OCCURRENCE: Perilla member, division 2b, of Lowell formation, southeastern Arizona. Types were collected in Guadalupe Canyon.

Groups of *Trigonia v-scripta* and *Trigonia vau*

The first of these two groups was established by Kitchin (1903, p. 12, 65-67) for species of British India like *Trigonia v-scripta* Kitchin, *T. dubia* Kitchin, and *T. recurva* Kitchin, in which the shell is compressed and elongated posteriorly, the beaks are placed at a considerable distance from the anterior margin, the areal ornamentation is obsolete, and the ribs have a peculiar V-shaped pattern in the more anterior part of the flank but become obsolescent posteriorly. At first Kitchin was inclined to unite these species with the group of *Trigonia vau*, but eventually decided to retain the too groups separately because of disagreement in the youthful development of the species respectively assigned to them.

The essential difference appears to be in the fact that in the *v-scripta* group both the marginal and inner carinae are clearly indicated, but there is no distinct separation of the narrow area into the inner and the outer parts. In the *vau* group, on the other hand, while a division into the inner and outer areas is definitely established in the mature stage, no carinae develop except the marginal carina, and even that is prominent only in the umbonal region. In the last group, instead of sharply pronounced median and inner carinae, there is (1) a longitudinal groove which divides the two areas, and (2) an obtuse angle which separates the area from the escutcheon. It seems to me that morph-

---

[22] Regarding the name on the label Dr. Reeside commented (personal communication, August 19, 1937): "The identification goes back a number of years and, I assume, was made by Dr. Stanton. It will be very useful, if you can show a definite sequence of recognizable forms in your area, for I think we now lack such knowledge. *T. taffi*, s.s., will have to rest on Cragin's cotypes (his lot 214 of which you have casts)."

ologically the *groove* in question is the median furrow which takes the place of the median carina in certain species of Trigoniae, and that the *obtuse angle* corresponds to the inner carina. Assuming the correctness of this interpretation, it is the absence of the median carina or the median furrow, and the accentuation of the inner carina, that essentially separates the species of the *v-scripta* group from those of the *vau* group.

It is the variability of the characters observed in the species of the *v-scripta* group that renders this entire group rather unstable. In *Trigonia recurva* Kitchin (1903, p. 75, Pl. 8, figs. 4–6), for instance, posteriorly the area is separated from the escutcheon not by a carina but by a furrow, which in one case seems to replace the carina altogether already in the anterior part (Kitchin, 1903, Pl. 8, fig. 4a). Resemblance of this species to *Trigonia vau* Sharpe was acknowledged by Kitchin himself. However, he found various characteristics of the two species, the absence of the inner carina in the entire *vau* group among them, as sufficiently warranting the retention of *T. recurva* in the *v-scripta* group.

Our present knowledge of these two groups of Trigoniae seems to justify the separation propounded by Kitchin. Yet the species of both groups have so much in common, and this refers not only to the V-shaped ornamentation of flanks, that the cognizance of their relation is imperative even in a broad discussion of separate members. As an instance, Kitchin (1926, p. 461) commenting on *Trigonia calderoni* (Castillo and Aguilera) from Malone, Texas, as identified and described by Cragin (1905, p. 59, Pl. 9, figs. 4–6), had to refer jointly to the *v-scripta* and *vau* groups. The features common to both groups are well established. The characteristic which is supposed to separate them widely—the relation between the areas and the carinae—is basically common to both and may be formulated as tendency toward the reduction of the median carina. This is complete in the *v-scripta* group. In the *vau* group the median furrow is present.

The two species described in this paper, *Trigonia cragini*, sp. nov., and *Trigonia kitchini*, sp. nov., are closer to the *vau* group and to those species of the *v-scripta* group which, like *T. dubia* and *T. recurva*, approach the *vau* group. I also place in this group *Trigonia conocardiiformis* (Krauss) from Africa as described and illustrated by Kitchin (1908, p. 119, Pl. 7, figs. 2–4). It is true that in this species the ribs of the anterior set do not fully attain a horizontal attitude in the anterior part of the flank as they do in these Arizona species and in "*Trigonia calderoni*" of Texas, to say nothing of the strongly V-shaped ribbing of *Trigonia vau* (Kitchin, 1908, Pl. 6, fig. 2). In mature stage of *T. conocardiiformis* from Uitenhage and Sunday's River (Kitchin, 1908, Pl. 7, figs. 2 and 4) only the ribs of a very broad and incipient V-pattern are present, but in the young stage (Kitchin, 1908, Pl. 7, figs. 2a–2b) this species is almost identical with *T. cragini*, a species in which the V-shaped ornamentation is fully developed in the later stages.

Probably the species such as "*T. calderoni*," *T. cragini*, *T. kitchini*, and *T. conocardiiformis* should be separated into a special group in which the pattern of the ribs does not form narrow V-s as it does in *T. vau* and *T. v-scripta*. In these forms the ribs of the anterior and posterior sets tend to meet at a right or at a wider angle. At present too little is known of the early development of the American species and of their geological relation to the groups of the southern realm. *Trigonia heterosculpta* Stanton (1901, p. 20, Pl. 4, figs. 16–18) from Patagonia has a variable ornamentation in which an outstanding feature is that the second anterior rib of the posterior set completely isolates it from the anterior set, a characteristic not observed in the *v-scripta* and *vau* groups.

Trigonia cragini Stoyanow, sp. nov.

(Plate 13, figures 6–10; Plate 14, figure 3)

Shell small, navicular in shape, with the greatest thickness in the anterior third, rapidly compressed posteriorly. Anterior margin is a broad, almost circular curve from the beak to the ventral margin which passes very gradually into the narrow and somewhat truncated posterior margin. Dorsal margin is concave.

The early ornamentation consists of about eight concentric umbonal ribs somewhat crowded near the anterior margin but spreading out toward the marginal carina which at this stage is a narrow and acute ridge. The first four ribs form a rather wide angle as they cross the carina but the rest of the initial ribs make an acute angle and normal V-s already on the narrow undifferentiated area, not sharply separated from the escutcheon, and converge under the beak. This condition continues for about 7 millimeters from the tip of the beak to the ninth rib. At this stage, the marginal carina suddenly becomes obtuse, and a median furrow, delineating a narrow outer area, appears. From this point posteriorly (a) the marginal carina, (b) the outer area, and (c) the median furrow are distinctly differentiated. Individualization of the inner area begins later, about the midlength of the shell, through the appearance of a longitudinal depression, or inner furrow, which rather inadequately separates it from the escutcheon proper.[23]

Ornamentation on the flanks changes as soon as the marginal carina loses its ridge-like character. From this point the ribs are composed of two sets: (1) the posterior set of vertical or subvertical ribs which are somewhat crowded near the marginal carina, at the contact with which they form small inverted V-s, but spread more widely in the direction of the ventral margin, and (2) the anterior set of horizontal or slightly curved ribs which turn a little upward only near the anterior margin. The ribs of these two sets (generally there is one rib of the horizontal set for each rib of the vertical set, but occasionally a few odd ribs may be present) meet posterior to the line of height to form acute but wide V-s so that the said line is nearly parallel with the ribs of the vertical set and normal to the ribs of the horizontal set. This is in a striking contrast to the Trigoniae of the *v-scripta* and *vau* groups in which the line of height either bisects the V-s, or is very near the bisecting line in the umbonal region.

Ornamentation of the escutcheon is composed of densely set costellae which at first cross the escutcheon diagonally but posteriorly spread over it in a fan-shaped way, being subparallel to the cardinal margin in its vicinity and set at an angle to the outer area. In the anterior part of the shell the costellae stop short at the median furrow and the outer area. In the posterior part they cease at the inner furrow and the inner area.

The youthful development of *T. cragini* is similar to that of *T. conocardiiformis* (Krauss) (compare Kitchin, 1908, p. 119, Pl. 7, figs. 2a–2b) in which the ribs on the umbo are set closer and make a more acute angle directly on the marginal carina. In the adult stage, however, the difference is strong. In the African species the ribs of the vertical set, crowded near the marginal carina, either form very incomplete, incipient V-s on the flank with the ribs of the anterior set, or simply pass into the latter with a broad forward sweep (Kitchin, 1908, Pl. 7, figs. 2, 3, 4). From the incomplete specimens illustrated by Kitchin (the posterior part of the shells missing) it cannot be inferred whether the individualization of the inner carina takes place in *T. conocardiiformis*. The older illustrations of Krauss are not available to the writer. The reproductions in Lycett (1879, p. 210–211, woodcuts) and the affinitive forms from South America illustrated by Burckhardt (1903, p. 72, Pl. 13, figs. 1–5) are not helpful in this connection.[24]

In the outline, size of the shell, and ornamentation of the flank, *Trigonia cragini* is very close to *Trigonia calderoni* (Castillo and Aguilera, 1895, p. 9, Pl. 5, figs. 17–18) from Catorce, San Luis Potosí, Mexico. Since, however, no description or illustration of the umbonal region of the shell, areas, and escutcheon was given by Castillo and Aguilera, the relation between these two species cannot be discussed. A more or less similar Trigonia from Malone, Texas, which may or may not be conspecific with the Catorce species, was identified and described by Cragin (1905, p. 59, Pl. 9, figs. 4–6) as *Trigonia calderoni*. Cragin did not discuss the escutcheon of the Malone form except stating that it was plain. I have examined Cragin's illustrated types which are deposited in the U. S. National Museum. These specimens are badly eroded, the beaks, areas, and escutcheon are destroyed so completely that it is impossible to make an interpretation as to their original nature. Dr. Reeside has kindly sent me a cast of the specimen collected at Malone by Dr. Stanton. The accompanying label reads:

---

[23] A perfect specimen showing the entire dorsal part of the shell was inadvertently lost during a demonstration at a gelological meeting. Two other specimens are substituted in which the umbonal and areal parts are seen separately (Pl. 13, fig. 6; and Pl. 14, fig. 3). The relation between the marginal carina, outer area, median furrow, inner area, inner furrow, and escutcheon in the posterior part of the shell is shown in Fig. 10 of Pl. 13.

[24] The relation between these African and South American forms has been discussed by Weaver (1931, p. 261–263). The entirely and uniformly undifferentiated area in *T. picunensis* Weaver (1931, Pl. 25, figs. 131–136; Pl. 27, fig. 150), with which Burckhardt's specimens were compared, is very remote from the areal development in *T. conocardiiformis* with the early individualized marginal carina and median furrow.

> *Trigonia calderoni* (C. & A.) of Cragin
> Malone formation of Cragin
> West base of Truncate Mound 30 feet beneath base of conglomerate, 1½ miles east of Malone Station, Texas
> Coll. 14263 (Stanton)

The umbonal ribs, the areas, the median and inner furrows are satisfactorily preserved in this homoeotype and are almost identical with those described in *T. cragini*, except that the outer area has fine longitudinal striae. The ornamentation of a rather narrow escutcheon either is not preserved, or the latter is, as Cragin postulated it, smooth. I have illustrated the cast in Figure 1-2, Plate 14, of this paper.

The Arizona species is very close to this form from Malone An appreciable difference is seen in the nature of the ribs of the horizontal set on the flank. In the latter form they are more widely spread than either in *T. cragini* or in "*Goniomya calderoni*" of Castillo and Aguilera, also possibly the beaks (partly destroyed) are placed less anteriorly. In the Malone form the angulation of ribs begins earlier and in such a way that the incipient normal V-s are entirely on the umbo, which results in early development of the V-shaped sculpture on the flank, whereas the marginal carina bisects the smaller inverted V-s. In *T. conocardiiformis* the normal V-s are on the marginal carina, and in *T. cragini* they pass early upon the flank. In the Malone form the ribs apparently do not extend from the marginal carina to the escutcheon. Of course, if the escutcheon in Cragin's specimens is actually smooth this would make an important distinction. However, the similarity is very considerable, and if the Arizona species were found in the same or equivalent strata with the homoeotype from Malone, the difference (barring the possibility of a smooth escutcheon in the Texas form) would most probably be considered of less than varietal value. Assuming Jurassic age of the Malone beds, the similarity between the Malone Trigonia and the Upper Aptian species of Arizona is truly remarkable.

TYPE: Syntypes: Nos. 91429, 91430, 91434, 91445, 91509.

OCCURRENCE: All syntypes were collected from Cholla member, division 6a, d, and f, of Lowell formation, Ninety One Hills.

Trigonia kitchini Stoyanow, sp. nov.

(Plate 14, figures 4-10)

Shell of medium size, high, comparatively short, inflated anteriorly, with the greatest thickness at the midflank but gradually decreasing to the posterior margin. Ornamentation, although of the same general type as in *Trigonia cragini*, is much more differentiated. On the excavated and depressed escutcheon it is quite peculiar. In the anterior part of the latter the costellae are as in *T. cragini*. Posteriorly, however, those near the inner area encroach upon it and merge with the lines of growth, but those in the middle part of the escutcheon and near the cardinal margin turn abruptly at a right angle and form a series of V-shaped zigzags, an escutcheonal ornamentation not known to exist in the *v-scripta* and *vau* groups.

The interrelation of the marginal carina, areas, median and inner furrows is essentially as in *T. cragini*. On the flank the ornamentation is also of the same general plan as in that species except that in the posterior part of the shell the last ribs of the vertical set turn abruptly backward at their dorsal termination and form well-sculptured short inverted V-s which soon merge with the lines of growth as they approach the outer area.

This species has a larger, higher, and relatively shorter and thicker shell than in *T. cragini* and *T. calderoni*. It differs essentially from these species in its more elaborate ornamentation. Besides the V-shaped pattern on the main portion of the flank, as in those species, it has V-s on the escutcheon and on the dorsoposterior part of the flank.

TYPE: Syntypes: Nos. 91499, 91505, 91506, 91507, 91603.

OCCURRENCE: All syntypes were collected from Cholla member, division 6a, of Lowell formation, Ninety One Hills.

Group of *Trigonia abrupta*

Affiliation of *Trigonia reesidei*, sp. nov., described below, with any of the established groups of Trigoniae presents considerable difficulties. The strongly trigonal outline of the shell and the obtuse marginal angulation occur in certain species of the Gibbosae, like *Trigonia lingonensis* Dumortier (*in* Lycett, 1879, Pl. 22, figs. 1-4). The peculiar arrangement of the riblets on the escutcheon and the areas, in which the riblets of the right and left valves are oriented obliquely outward from the cardinal margin to the marginal carinae in such a way as to form broad normal V-s, is not observed in the majority of described Trigoniae ornamented with oblique riblets In such species the riblets of the right and left valves almost invariably form inverted V-s. The normal V-s within the marginal carinae of the entire shell occur occasionally in very remote groups, as for instance, in the Scabrae and in some of their derivatives, like *T. aliformis* Parkinson (*in* Lycett, 1879, Pl. 25, fig. 4a), *T. sulcataria* Lamarck (*in* d'Orbigny, 1843, Pl. 294, fig. 6), and in *T. rogersi* Kitchin (1908, Pl. 4, fig. 1), which in no way are related to the described species from Arizona. The rectangular V-shaped pattern of the ribs on the flank is preeminently developed in the above discussed representatives of the *v-scripta* and *vau* groups which likewise have very little in common with *T. reesidei*. It may be argued whether the *abrupta* group was derived from the Undulatae, exemplified by *T. undulata* Fromherz (*in* Agassiz, 1840, Pl. 10, figs. 14-16), through the reduction and rounding of the truncate posterodorsal margin, and the shifting of the marginal carina dorsally, with the consequent greater differentiation of the antero-ventral region of the shell, or whether the apparent similarity is a matter of convergence.

Little is known about the group connections of species like *Trigonia abrupta* L. v. Buch (1839, p. 17, Pl. 2, figs. 21-22), *T. humboldti* L. v. Buch (1839, p. 9, figs. 28-30), and *T. lorentii* Dana (1849, p. 721, Pl. 15, fig. 12) which undoubtedly are related to *T. reesidei*. Gillet (1924, p. 96) attempted to connect *T. abrupta* with *T. hondaana* Lea, a postulation considered untenable by Dietrich (1938?, p. 98). It is true that the areal riblets of one valve of this species, as illustrated by Coquand (1866, Pl. 24, figs. 1-2) and later by Gillet (1924, p. 95, fig. 51) are inclined to form wide V-s with the corresponding costellae of the opposite valve, but the shape and ornamentation of the entire shell altogether disagree with v. Buch's species. Better illustrations of Gerhardt (1897, Pl. 5, figs. 6a-6b) and Dietrich (1938?, Pl. 18, figs. 1-7) show a complex pattern of the areal ornamentation in *T. hondaana* which by the total sum of characters is close to the Quadratae, as was emphasized by Dietrich. For *T. abrupta* L. v. Buch Dietrich proposed subgenus *Buchotrigonia*.

The group Abruptae is proposed tentatively. A discussion of species which may be placed in it is given in the description of *Trigonia reesidei*.

Trigonia reesidei Stoyanow, sp. nov.

(Plate 14, figures 11-14; Plate 15, figures 1-3)

Shell moderately convex, trigonal when young, ovately trigonal in the adult stage. Average mature specimens are about 40 mm. long, 35 mm. high, and 25 mm. thick, with maximum inflation slightly above the midlength a little posterior to the line of height. Anterior margin, somewhat concave immediately in front of the umbonal region, slopes forward in the upper third of the shell, forms a gentle curve anteriorly, and gradually passes into the ventral margin. The foremost point in full-grown shells is a little below the midheight. Dorsal margin, slightly convex immediately behind the narrow, oval ligament pit, descends abruptly in an almost straight line to the short posterior margin which is well rounded in the adult stage. Posterior margin passes by a continuous line into the gently convex and regularly curved ventral margin which has the greatest convexity directly opposite the beaks.

Umbones are in the anterior half of the shell and placed slightly forward of the equidistant line between its foremost and hindmost points. They are produced into slender pointed beaks which are directed almost centrally, or are very little recurved. Flank is separated from the outer area by a marginal carina which is ridge-like in the young stage to the height of 15-18 mm. and passes posteriorly into and obtuse and broad marginal angulation that still distinctly separates the flank from the area to the posterior margin. In the young stage the outer area is gently inclined inward, toward the well-pronounced median furrow which separates it from the narrower and slightly elevated inner area. The inner area slopes toward the escutcheon from which it is separated by a not sharply outlined longitudinal depression, or the inner furrow, which is broader and shallower than the median furrow and dwindles posteriorly more rapidly than the latter.

Ornamentation consists first of a set of initial subconcentric ribs of the umbonal region, widely spread near the anterior margin of the shell but crowded toward the marginal carina. The very first ribs actually cross the marginal carina and are set at an angle to the initial riblets, thus forming small inverted V-s on the outer area. Since, however, the number of areal riblets increases very rapidly, out of proportion with the number of ribs, it is more proper to assume that the ribs of the initial set, as a whole, are at a contact with the riblets rather than pass into them. This initial set extends from the beaks for 10-15 mm. down the flanks and contains about 12-14 ribs. Gradually the crowded posterior parts of the ribs which are near the marginal carina become steeply inclined, in the same time their anterior subconcentric and subhorizontal parts develop small, closely spaced, zigzagging V-s. The initial set terminates at the indicated distance from the beaks, at which point the posterior parts of the ribs are already vertical and meet the subconcentric ribs at the right angle. In this way two sets of ribs, one horizontal and anterior, and the other vertical and posterior, are established on the main part of the flank. With the passing of subconcentric ribs into horizontal ones, farther down ventrally, the zigzagging pattern moves anteroventrally away from the middle part of the flank. Between the midflank and the ventral margin some of the ribs of the horizontal set, which here are concentric with the ventral margin, cross partly or entirely the ribs of the vertical set. With growth, the right angle between the vertical and horizontal ribs gradually passes toward the posteroventral termination of the shell and stops short of it near the obtuse terminal portion of the marginal angulation.

Ornamentation of the areas and the escutcheon is composed of a single set of closely spaced riblets. They begin at the cardinal margin on the escutcheon as short low ridges which very soon swing forward obliquely across both areas and the furrows to the marginal carina. There is an incipient tendency toward zigzagging on the escutcheon just before the inner area is reached. When both valves are in juxtaposition, the riblets of the right and left valves form densely set and wide normal V-s. The lines of growth on the flank are obscured by the strong ribs of the vertical set, but the latter evanesce in the posterior part of the shell, and the lines of growth become distinct as they cross the wide, posterior portion of the marginal angulation. Passing over the angulation they also cross the riblets at an angle, which causes a quincunx pattern in the posterior part of the areas and the escutcheon. In the posteroventral termination of the shell the marginal angulation and the areas bear only the lines of growth.

Trigonia abrupta L. v. Buch, as illustrated by Dietrich (1938?, Pl. 19, figs. 3, 4, 5a, 5b) is close to *Trigonia reesidei*. The ornamentation of this South American species does not show zigzagging ribs on the flank. Its area and escutcheon were neither described nor illustrated by Dietrich. Dietrich is of opinion that *T. humboldti* L. v. Buch is only a poorly preserved specimen of *T. abrupta*. However, the specimens of *T. humboldti* figured by Dietrich (1938?, Pl. 20, fig. 1 [holotype, Humboldt's collection] and Pl. 20, fig. 2 [Karsten's collection]) show that in *T. humboldti* the ribs of the anterior set are of closely undulating, or even zigzagging nature, a feature lacking in the forms identified by Dietrich with *T. abrupta* in which the broadly undulating or wavy ribs do not form short zigzags, nor are the zigzagging ribs seen in v. Buch's original drawing of the holotype of *T. abrupta* and in Lisson's illustration (Lisson, 1930, Pl. 10, fig. 1). *T. humboldti*, as represented by v. Buch's holotype and Karsten's plesiotype, apparently is distinguished from *T. abrupta* by the presence of zigzagging ribs in the anterior and anteroventral parts of the flank and by its different outline, though Humboldt's and Karsten's forms may not be conspecific.

Judging from v. Buch's original illustration, *T. abrupta* is a form very elongated in the anterodorsal-posteroventral direction. Its vertical posterior ribs reach the ventral margin, whereas the anterior horizontal ribs turn upward posteriorly and reach the marginal carina independently without abutting against the vertical ribs, except the three lower ribs which either are in contact with, or pass into, the first vertical rib. In v. Buch's species the horizontal ribs do not pass into the vertical set above the midflank; the ribs, as mentioned above, lack the zigzagging pattern; and the entire shell is much longer than in *T. reesidei*.

Trigonia abrupta from Spain, illustrated by Coquand (1865, Pl. 23, figs. 4–5) and identified by him with v. Buch's species, is a specimen in which 15 or 16 ribs of the anterior set pass into the ribs of the vertical set by forming wide and open V-s, except in the extreme ventral region where the posterior ribs of the vertical set almost reach the ventral border and have no counterparts in the anterior set of ribs. According to Lycett, Coquand's form is close to *T. meyeri* Lycett (1879, p. 126, 227, Pl. 23, fig. 6; Pl. 41, figs. 15-16). The scabroid shell, concave dorsal margin, and strongly developed marginal carina which extends to the posterior margin are against the affiliation of the *T. meyeri* with v. Buch's species.

In discussing *T. abrupta*, Dietrich (1938?, p. 98) made reference to *Trigonia lorentii* Dana (1849, Pl. 15, fig. 12). I have had an opportunity to examine Dana's original collection in the U. S. National Museum. In this collection the shortest shell measures: height 59 mm., length 69 mm.; and the longest specimen: height 39 mm., length 70 mm. The best preserved syntype in Dana's collection is relatively the longest specimen with a blunt angulation which extends from the beak to the posteroventral extremity and separates the flank from the areal region. The ornamentation of the flank does not pass on the areas which have only concentric striae. The outer area is separated from the inner area by a narrow and shallow median furrow which is quite well marked in the anterior and middle thirds, but is barely perceptible in the posterior third of the shell. The inner area has a feeble longitudinal elevation on the side of the median furrow, then a shallow groove followed by another slightly elevated fold which separates the areal region from the escutcheon proper. The escutcheon has no individual ornamentation and is crossed by striae which extend to the areas. Ornamentation of the flank consists of subvertical ribs of the posterior set and subhorizontal ribs of the anterior set. On the midflank and in the ventral region of the shell the ribs of two sets form acute V-s, which arrangement almost reaches the ventral margin. In the umbonal region the V-s are less acute. This part of the shell is somewhat corroded, but it is evident that there were no zigzags in the horizontal ribs, although they undulate somewhat between the vertical set and the anterior margin. The axis which bisects the V-s is not a straight line but roughly converges with the marginal carina toward the beak.

Trigonia lorentii Dana as illustrated by Lisson (1907, Pl. 3, figs. 1a–1c, 3) has horizontal and vertical ribs that meet individually almost as far as the ventral margin. The V-angle is more acute than in *T. reesidei*.

Of all discussed Trigoniae the specimen designated by Karsten as *Trigonia humboldti* and illustrated by Dietrich (1938?, Pl. 20, fig. 2), of which only the anterior part is preserved, alone has ornamentation on the flank similar to that of *T. reesidei*, i.e., the zigzagging ribs already indicated in the umbonal region and well developed in the anterior part of the flank. However, these two Trigoniae are not conspecific. The anterior margin of Karsten's specimen is a continuous curve from the beak to the ventral margin, in a striking contrast to *T. reesidei* in which the umbonal region is well sculptured, and the anterior margin, as described, abruptly slopes forward toward the foremost point of the shell wherefrom it gradually recedes to the ventral margin.

Among North American Trigoniae there are three species that require discussion. *Trigonia conradi* Meek and Hayden (1864, p. 83, Pl. 3, fig. 11), from the southwest base of the Black Hills found in the Upper Jurassic beds of that region, may belong to the Abruptae. According to the original description:

"The specimens ... are not in a condition to have retained fine surface markings if there were any ... The escutcheon was not satisfactorily preserved though ... it seems to have been marked by obscure radiating costae, and is bounded on each side by the distinctly angular umbonal slopes...."

I have examined the type preserved in the U. S. National Museum. It is marked as "Type No. 212." The type specimen does not show the anterior subhorizontal and the posterior subvertical ribs connect on the flank as Meek's figure 11 in Plate 3, of the 1864 paper indicates. The middle part of the flank is corroded and no ornamentation is observable there. Another specimen, in the same box with the type, shows fine riblets encroaching over a blunt marginal carina on the flank in the posterior half of the shell. This may be a different species.

Two undescribed species in the Cretaceous strata of Texas undoubtedly belong in the Abruptae. The specimens are deposited in the U. S. National Museum with the following label:

Trigonia sp. 1
Travis Peak form
Loc. 1804
Panther Creek, on road from Travis Peak to Burnet,
between 18th and 19th mile posts from Burnet.

Through the kindness of Dr. Reeside I have casts of the three specimens of this set. The largest specimen is probably identical with *Trigonia abrupta* as illustrated by Dietrich (1938?, Pl. 19, fig. 3). Another specimen, slightly shorter and longer, is most probably conspecific with the first. Both specimens have a smooth areal region in which only the lines of growth are present. The third and smallest specimen, however, has undulating horizontal ribs in the anterior part of the flanks and, what is very interesting, has riblets that form inverted V-s in the anterior part of the escutcheon. This is a different species from *T. abrupta* and may be close to *T. reesidei* in which, as stated above, the costellae show incipient zigzagging as they pass from the escutcheon onto the inner area.

TYPE: Holotype: No. 91049. Paratypes: Nos. 91050, 91051.

OCCURRENCE: Abundantly represented in lower part of Lowell formation from its base in Lancha limestone, division 9g, Pacheta member, where illustrated types were collected, through Espinal grit, division 8c, Joserita member, to Arkill limestone, division 7l, Savedra member. Ninety One Hills and Black Knob Hill.

Group of *Trigonia Excentrica*

The *excentrica* group was suggested by Lycett (1879, p. 7) as a part of Agassiz's section the Glabrae. Later Bigot (1892, p. 268. *Compare* Kitchin, 1903, p. 9) separated the entire section of Agassiz into the Semilaeves, the Gibbosae, and the Excentricae. The *excentrica* group, which contains a limited number of species, is characterized, in common with the *gibbosa* group, by concentric ribs which on the flank do not reach the marginal carina and the posterior margin. This arrangement produces a smooth space nearly parallel to the marginal carina from the umbonal region to the posteroventral extremity. As a result, the ornamentation is accentuated in the anterior and middle parts of the flank. The unornamented ribbing on the flank, greater reduction of the carinae, and simple, little ornamented nature of the areas and escutcheon separate the species of the *excentrica* group from those placed in the *gibbosa* group.

To the *excentrica* group, most probably, belongs *Trigonia lerchi* (Hill, 1893, p. 30, Pl. 4, fig. 3; *compare* Gillet, 1924, p. 83) found in the "heavy" conglomerate at the base of the Comanche series on Sycamore Creek, near the crossing of the Burnet and Travis Peak roads, Burnet County, Texas.

Dr. Reeside very kindly sent me casts of two specimens, on deposit in the U. S. National Museum, apparently from the same or a neighboring locality, which undoubtedly belong in the *excentrica* group, and one of which seems to be related to *Trigonia weaveri*, sp. nov., described below. The two specimens represented by these casts are not conspecific. The casts were accompanied by the following label:

Trigonia sp. 2
Travis Peak

Loc. 1804
Panther Creek, on road from
Travis Peak to Burnet, between
18th and 19th mile posts
from Burnet.

Evidently these examples were collected together with the Texas forms of the *abrupta* group discussed in this paper.

Trigonia weaveri from the Lowell formation, a new species of the *excentrica* group, is close to *Trigonia agrioensis* described by Weaver (1931, p. 266, Pl. 27, figs. 142–146) from the Cretaceous strata of Argentina.

Trigonia weaveri Stoyanow, sp. nov.

(Plate 15, figures 4–8)

Shell of medium size, considerably inflated in two anterior thirds, produced and attentuated posteriorly. Anterior margin at first gently slopes forward from the beaks in the umbonal region, but before reaching the anterior extremity of the shell, which is at its midheight, passes into an elliptical curve continuous with the ventral margin to the posteroventral region, from there the curve rises more steeply to the posterodorsal extremity situated above the midheight of the shell. Dorsal margin is convex in the umbonal region and gently concave posteriorly.

The entire outline of the shell is irregularly ovate with the umbonal region well sculptured and the posterodorsal extremity (apparently) truncated above by the dorsal margin. Slightly recurved beaks are placed but little anterior to the midlength of the shell.

Ornamentation of the umbonal region consists of concentric ribs of a long radius. In some specimens these ribs are crossed by a low, narrow, and ill-defined ridge. The latter extends from the beaks at an acute angle with the anterior margin and delineates a triangular space in front. It dwindles ventrally and ceases on the flank before the midheight of the shell is reached. In some specimens this ridge is not developed, and in others it is placed quite a distance from the anterior margin. Posteriorly, the umbonal ribs cross the marginal angulation which separates the flank from the areas and, sharply turning inward, encroach upon the areas under the beak and reach the cardinal margin.

Ornamentation of the flank proper is composed of concentric ribs which become nearly horizontal in the middle part of the valve. The ribs are somewhat wavy in the anterior part of the flank as they approach the anterior margin. This, however, is not a constant character, as variations from almost straight ribs to those possessing an appreciable sinuosity are observed. The "excentric" nature of ornamentation is strongly accentuated in the anterior part of the shell where the ribs cross the lines of growth, as in *T. excentrica* Parkinson (Lycett, 1879, Pl. 28, fig. 6). Posteriorly the ribs dwindle about the midlength of the shell in the dorsal part of the flank, near the umbonal region, but become progressively longer in the ventral region.

The angulation that separates the flank from the areal region apparently represents the marginal carina. The entire areal region, with the narrow, relatively elevated and steeply inclined areas, and with the wider, depressed and excavated escutcheon, slopes toward the cardinal margin. The narrower outer area is fairly well defined between the marginal angulation above and the median furrow below with which it merges. The wider inner area is ill-defined on the side of the escutcheon, yet it is easily distinguished from the latter by the presence of a shallow longitudinal depression which inwardly merges with the escutcheon, but especially by the fact that the ribs which encroach from the flank on the outer area also cross the inner area, especially in its anterior part, but do not extend to the escutcheon.

T. weaveri differs from the well known species of the *excentrica* group, like *T. excentrica* Parkinson (Lycett, 1879, p. 94, Pl. 20, figs. 5–6; Pl. 21, figs. 6–7; Pl. 22, figs. 5–5a; Pl. 28, figs. 6–6a, 9–10) and *T. dunscombensis* Lycett (1879, p. 188, Pl. 40, figs. 5–6) in its more attenuated posterior region and more concave dorsal margin. *T. laeviuscula* Lycett (1879, p. 96, Pl. 22, fig. 6) disagrees with the Arizona species in its narrow umbonal region, more medially placed beaks, and broader posterior extremity.[25]

T. weaveri is close to *T. agrioensis* Weaver (1931, p. 266, Pl. 27, figs. 142–146). *T. agrioensis* differs in being relatively longer. The wavy character of the ribs in the anterior region of its shell is much less pronounced. The ridge which crosses the ribs on the umbo is closer to the anterior margin. However, as in *T. weaveri*, the development of this ridge is variable in Weaver's species. It is quite strong in the form illustrated in Figure 142, Pl. 27, of Weaver's monograph, less pronounced in the example in Figure 146 of the same plate, whereas Figure 144 shows a specimen with barely perceptible ridge. In both species the ribs may not develop between this ridge and the anterior margin.

TYPE: Syntypes: Nos. 91997, 91200, 91245. Metatypes (not illustrated): Nos. 91033, 91201, 91246, 91350, 91440, 91806, 91812, 91814, 91817, 91819, 91938, 91998, 91999, 92018.

---

[25] The areal view of these species is not adequately represented in Lycett's illustrations.

OCCURRENCE: Abundant in greenish shale and limestone of Quajote member, division 3d, Lowell formation. Trigoniae of the *excentrica* group, not satisfactorily enough preserved for detailed studies, have also been collected from Espinal grit, Joserita member, division 8c, and from Arkill limestone, Saavedra member, division 7l. Ninety One Hills.

Group of *Trigonia aliformis*

In this group Lycett (1879, p. 8, 115, 217) placed (as a sub-group) the species of the Scabrae with an inflated anterior region and with a marked elongation and attenuation of the posterior region of the shell. Species within the group range from those in which this attenuation reaches extreme proportions, as in *T. aliformis* Lamarck and *T. caudata* Agassiz (*see* Lycett, 1879, Pl. 25, figs. 3–6; Pl. 28, figs. 5–5a; Pl. 26, figs. 5–6b), and the areas, with a reduced ornamentation or unornamented, are narrow, to those grading into the Scabrae in which the posterior elongation is at its minimum, the thickness of the shell decreases gradually toward the posterior margin, and more strongly ornamented areas are considerably wider, as in *T. crenulata* Lamarck (*see* d'Orbigny, 1847, Pl. 295, figs. 1–4). Both extreme types of the group are represented in the Cretaceous strata of Arizona.

In the rather outstanding ventricose representatives of this group, like *T. ventricosa* (Krauss, in Lycett, 1879, p. 119, woodcuts; Kitchin, 1903, Pl. 10, fig. 5) of the Indo-African Province, the marginal carina is well developed from the beaks to the posterior region. The relatively broad, elevated outer and inner areas are separated by the median furrow, and the inner area is differentiated from the escutcheon by the presence of a longitudinal depression (Kitchin, 1903, p. 105, interpreted the inner elevation as a part of the escutcheon), both areas are ornamented in the entire anterior half of the shell.

In the typical species of the *aliformis* group the marginal carina is indicated only in the umbonal region. It grades more or less early into a marginal angulation which posteriorly merges with the rapidly expanding outer area. Both areas are adequately differentiated, but they are very narrow anteriorly and in forms with reduced areal ornamentation effectively separate the riblets of the escutcheon from the ribs of the flank. (*See* Lycett, 1879, Pl. 25, fig. 4a; Pl. 26, fig. 6b; Pl. 27, fig. 1a; Pl. 28, fig. 5a.)

In the typical Scabrae (d'Orbigny, 1843, Pl. 296, fig. 2) on the other hand, the areas are less individualized, but the strong ornamentation of areal riblets, continuous with those of the escutcheon and with the ribs of the flank, extends to their posterior termination.

Trigonia stolleyi Hill

(Plate 15, figures 9–11)

1893. *Trigonia stolleyi* HILL, The paleontology of the Cretaceous formations of Texas: The invertebrate paleontology of the Trinity division, Proceedings of the Biological Society of Washington, vol. 8, p. 26, Pl. 3, figs. 3 and 5.

Hill's original diagnosis of this species is essentially as below:

"Semi-lunate in general outline, beaks well forward and strongly recurved; anterior and pallial margin a strong continuous curve; posterior portion elongated with truncated posterior margin; cardinal area compressed. Surface marked by flexuous noduled costae, about twenty-two in number, narrow and high, separated by broad intercostal areas as in *T. alaeformis* Lmk.; depressed cardinal area bordered on its outer side by a long, narrow groove and marked by cross-ribs flexing anteriorly."

To this the following additional description may be added:

Shell relatively short, considerably inflated in the anterior half, thickest slightly posterior to the umbonal region, and becoming gradually compressed toward the posterior margin. The curve comprising the anterior, ventral, and posterior margins, from the beaks to the posterodorsal extremity, may be described as two involutes of the circle which intersect at the greatest depth of the ventral margin about the midlength of the shell. The anterior involute, of a circle of greater radius, abruptly undercuts the shell anteriorly, whereas the posterior margin is formed by a more sharply curving involute. Consequently, the anterior margin is very short. Dorsal margin is concave, steeper in the umbonal region and slightly rising posteriorly.

Flanks curve gradually from the dorsal to the posteroventral margin in the posterior part, but in the anterior half of the shell they sharply bend about its midheight toward the anterior margin and thus form obtuse but prominent angulation in the middle. As a result, the shell is very broad and flattened in front.

Beaks are strongly curved and slightly directed backward. Areal region is oval, little differentiated anteriorly. Areas are narrow and unornamented. Marginal carina is narrow and linear in the umbonal region. It early passes into the marginal angulation which posteriorly dwindles and merges with the elevated outer area. The latter is separated from the inner area by the deep and narrow median furrow. The inner area, about the same width as the outer area, is separated from the escutcheon by an indefinite longitudinal depression. Ligament pit either is partly inter-umbonal or is placed immediately behind the beaks.

Ornamentation of the umbonal region consists of prominent sharp subconcentric ribs which are rather widely separated on the umbo but converge both toward the undifferentiated areal region and escutcheon under the beaks and toward the anterior margin, without reaching the latter. On the flank, the ribs are slightly flexuous with a shallow forward concavity. In the dorsal part of the flank they converge in a general way toward the point of differentiation of the areas and abut against the marginal angulation. Ventrally the ribs considerably increase in height and width, and in the anterior part of the flank they reach the anterior margin.

The ribs cross the marginal carina and connect with the escutcheonal riblets only in the extreme umbonal region where the areas are not differentiated. The posteriorly differentiated outer and inner areas, and the median furrow between them, are crossed by fine straight, distantly spaced striae of a greater number than the ribs, and separate the escutcheonal ornamentation from that of the flank. On the escutcheon the coarse, rounded riblets begin from the cardinal margin as shallow wide inverted V-s, but become straighter, sometimes subhorizontal, or even shallowly concave in the middle, and slightly bend anteriorly as they approach the inner area. The ribs are crossed by prominent longitudinal nodes, as has been pointed out by Hill. Riblets on the escutcheon have similar but more delicate nodes.

The shell of this species is rather short for the *aliformis* group. However, the early reduction of the marginal carina and the absence of the ribs or riblets on the narrow areas, except in the extreme umbonal region, place *T. stolleyi* closer to this group than to any other group of the Scabrae.

The Arizona specimens are identical with Hill's holotype which I have examined in the U. S. National Museum.

PLESIOTYPES: Nos. GU20, GU33.

OCCURRENCE: Abundant in Perilla member of Lowell formation where it is found with *Trigonia mearnsi*. Varietal forms known as low as Saavedra member. Illustrated specimens are from the Guadalupe Canyon.

Trigonia sp. ex *aliformis* group

(Plate 16, figures 1–3)

The shell of this form, much larger than that of *T. stolleyi*, is inflated in the anterior part and very caudate and narrow posteriorly. Anteroventral margin is nearly a continuous curve. Posterior margin turns steeply and obliquely upward to the posterodorsal extremity in an almost straight line. Dorsal margin gently slopes from umbonal region, forms a broad concavity in the middle, and rises again posteriorly. Ribs are more numerous and more flexed in ventral region than in *T. stolleyi*. Details of areas and escutcheon are unknown.

In its general habitus this form is fairly close to the caudate species of the *aliformis* group, like *T. aliformis* Parkinson and *T. caudata* Agassiz, including such special features as the strong separation of the areas caused by the deepening of the median furrow in the posterior region. (*Compare* Lycett, 1879, Pl. 25, figs. 3–6; Pl. 26, figs. 5–6.)

I have no fully preserved specimen of this species at the time of this writing, although good examples may undoubtedly be obtained in the Guadalupe Canyon. Internal molds are not unlike the badly worn specimen from the bluffs of the Colorado River near Bull Creek, Travis County, Texas, identified and illustrated by Hill (1893, p. 27, Pl. 3, fig. 4) as *T. crenulata* Roemer. However, in the described form the ribs are wider and more flexuous in the ventral region and the posteroventral margin rises more abruptly. Independently of whether or not Hill's specimen is identical with the form from Fredericksburg referred by Roemer (1852, p. 51, Pl. 7, fig. 6) to *Trigonia crenulata* Lamarck (*see* d'Orbigny, 1847, Pl. 295), the Arizona examples belong to a species different from the latter Trigonia. They also are sufficiently distinct from *Trigonia emoryi* Conrad (1857, p. 148, Pl. 3, figs. 2a–c) with which Böse (1910, p. 121; Adkins, 1928, p. 121) compared Hill's specimen. The South American *Trigonia tocaimaana* Lea (*in* Dietrich, 1938?, Pl. 19, figs. 1–2b) is less caudate and its posterodorsal extremity is less undercut.

SPECIMENS: Nos. 91740 from Ninety One Hills and GU72 from the Guadalupe Canyon.

OCCURRENCE: Perilla member of Lowell formation together with *T. mearnsi* and *T. stolleyi*.

Family PECTINIDAE Lamarck

Pecten Osbeck, 1765

Pecten (Chlamys) thompsoni Stoyanow, sp. nov.

(Plate 16, figures 5–8; Plate 17, figure 1)

1904. *Pecten stantoni* STANTON (non HILL), *in* Ransome. The geology and ore deposits of the Bisbee quadrangle, Arizona, U. S. Geol. Survey, Prof. Paper 21, p. 67 and 70.

This species varies considerably in the proportions of height and length, in the convexity of the shell, and to a lesser degree in the details of ornamentation. In the average specimens the shell is ovate and subequilateral, with the anterodorsal margin slightly concave, posterodorsal margin a little convex or straight, and ventral margin gently rounded. Ears are of medium size and very unequal.

Right valve is from flattened to somewhat convex in the middle, left valve is appreciably more convex with the convexity gradually increasing from the anterior, posterior, and ventral margins and attaining its maximum in the dorsal half of the valve.

Ornamentation in the middle of each valve consists of primary ribs, moderately rounded and somewhat flattened on the top, which regularly alternate with much lower secondary ribs set so closely that the interspaces appear as very narrow but relatively deep grooves. Toward the anterior border the primary ribs become slightly split longitudinally by a very shallow groove so as to form a compound rib made up of two riblets, a third additional and more individualized riblet soon appears on the posterior side of the primary rib, and the entire compound primary rib forms then a "triplet." More anteriorly, the primary ribs again become simple and differ from the secondary ribs only in the size. Near the anterior margin the primary and secondary ribs are alike.

Similar modification in the ribbing is observed in the posterior part of the valves with the difference that when a primary rib becomes a "triplet" the tertiary rib develops on its anterior side. The ribs bear transverse lappet-like scaly projections, the concentric arrangement of which is more accentuated in the anterior and posterior parts of the shell.

With age the primary ribs become compound closer to the median part of the shell. The ribs of the right valve become flatter, those of the left valve more elevated. In the ventral region of mature specimens, and especially in the left valve, the primary ribs lose their compound nature and appear as single solid ribs. In large examples the ribbing often is obsolete in the posteroventral region. The lappet-like projections also lose their individual character and aggregate into continuous ridges concentric with the ventral margin.

The anterior ear is the larger, and in young specimens is quite long relative to the size of the shell. The byssal sinus, with a strong ctenolium, is well developed. The posterior ear is of a trigonal outline. There are only incipient, barely perceptible ribs on the ears.

Pecten thompsoni is related to the group of European Cretaceous pectens exemplified by *Pecten elongatus* Larmarck as illustrated by Woods (1903, Pl. 31, figs. 10–13; Pl. 32, figs. 1–3). *Pecten (Chlamys) elongatus* is a species with a higher and more equilateral shell, more attenuated in the apical region, with narrower, more pronounced and elevated, and predominantly rounded ribs on the median portion of the valves; its compound ribs are much less regularly distributed, alternate with the simple ribs without any definite order, and are not as wide and flat as the compound ribs of the Arizona species.

Specimens labelled *Pecten stantoni* in Ransome's collection from Bisbee, now on deposit in the U. S. National Museum, belong in *Pecten thompsoni*. Stanton's sketch of "*Pecten stantoni* (Hill)" (*in* Ransome, 1904, Pl. 16, fig. 2) most probably was made from an incomplete specimen of *P. thompsoni*. The only form of the Bisbee sequence comparable at all to *Pecten stantoni* Hill (1893, Pl. 2, figs. 3-3a) is a small incomplete pecten, collected from the Barata limestone of the Saavedra member (D. 7n), which similarly has strong furrows on the flattened primary ribs. As both Hill's illustrated type and the Arizona specimen are incomplete, I was unable to compare these forms more closely.

TYPE: Syntypes: Nos. 91400, 91955, 91399, 91951.

OCCURRENCE: Abundant in Saavedra member, division 7d, of Lowell formation, where it is found together with *Ostrea edwilsoni*. Ninety One Hills.

Neithea Drouet, 1825

Neithea vicinalis Stoyanow, sp. nov.

(Plate 11, figures 6-7)

This small, subtrigonal and nearly equilateral shell, represented in my collection only by one specimen of right valve, belongs in that division of *Neithea quinquecostata* (Sowerby) group which includes the species with tertiary ribs and smooth areas.

Accordingly, the right valve has six prominent primary ribs with major interspaces between every two of them.

On the inward side of the anterior (the first) and posterior (the sixth), and on the outward side of the second and fifth ribs, there are strong tertiary ribs set close to the corresponding primaries though they remain as individual ribs to the very apical part of the shell. On the inward side of the second and fifth, and on both sides of the third and fourth primary ribs, the tertiary ribs are somewhat less individualized in the ventral part of the valve where they are more or less flat and appear to be composed of two or more longitudinal striae. Dorsally, however, they pass into somewhat stronger single ribs.

In the middle part of the valve, between the second and fifth primaries, there are two secondary ribs in the major interspaces between every two primaries. These ribs are narrow and strong in the apical part of the specimen but become wider and flatter ventrally. In the ventral part of the valve they are composed of from two to four longitudinal striae. Here, however, their appearance partially depends on the preservation of the shell surface. Anterior to the second, and posterior to the fifth primaries, the secondary ribs are well rounded and more elevated in the ventral region. The total number of secondary ribs is ten.

The deepest intercostal depressions, or interstices, observed on the specimen, are between and on the sides of the secondary ribs. They seem to be as deep as the secondary ribs are elevated but are narrower than the latter.

The ears are small and trigonal. On the anterior area there are very irregular longitudinal wrinkles which cannot be interpreted as ribs. The posterior area is entirely smooth.

Except for areas, the surface of the specimen is covered with fine, minute, and closely set striae which run concentrically with the margin of the valve.

In geological age, general appearance, and smooth nature of the areas, this species is closer to *Neithea morrisi* (Pictet and Renevier) from Hythe Beds of Lympne, England, (Woods, 1903, p. 201, Pl. 39, figs. 11a-11c) than to any other described European *Neithea*. From this form of the Lower Greensand the Arizona species differs in the more trigonal outline of the shell, more inflated right valve, and less prominent secondary and tertiary ribs. An examination of the original picture of *"Janira morrisi"* Pictet and Renevier (1858, Pl. 19, figs. 2a-2d) shows a form with a rather coarse costation. Gillet (1924, p. 53, fig. 30) mentions five [?] primary and four intermediate ribs in *Neithea morrisi* (Pictet and Renevier) although in the accompanying illustration there are three minor ribs between two primaries.

The only American species of the *Neithea quinquecostata* group with regular tertiary ribs and smooth areas are those originally described as *Pecten quadricostatus* var. Roemer (non Sowerby) (Roemer, 1852, p. 64, Pl. 8, figs. 4a-4c), *Neithea occidentalis* Conrad (1857, p. 150, Pl. 5, figs. 1a-1c), and *Vola irregularis* Böse (1910, p. 97, Pl. 15, figs. 10-18).[26]

---

[26] It is clear from description of Whitfield (1885, p. 56, Pl. 8, figs. 12-14) and S. Weller (1907, p. 481, Pl. 51, figs. 7-12), that *"Neithea quinquecostata* (Sowerby)" of the New Jersey Cretaceous has ornamented areas.

As seen from Roemer's illustrations, in *Neithea quadricostata* var. the general arrangement of ribs is of the same order as in *Neithea vicinalis*. The right valve of the Texan species, however, is notably attenuated in the apical portion, and its ribs diverge rather rapidly toward the ventral margin which gives the entire shell a subpentagonal and not trigonal outline. *Neithea occidentalis* Conrad, as illustrated by Meek, is remarkably similar to the Arizona form. It even seems that the secondaries between the third and fourth primary ribs are flattened in the same degree and also split into striae in the ventral region, and that the major interspace between those primary ribs is narrow, contrary to all other related species except *Neithea morrisi* (Pictet and Renevier, part) and *Neithea irregularis* Böse. Unfortunately, the type of *Neithea occidentalis* Conrad preserved in the U. S. National Museum is so different from Meek's illustration as to cause reasonable doubt that Meek had that specimen in view as an example for his drawing. *Pecten quadricostatus* var. Roemer has been identified with Conrad's species by Conrad (1855, p. 269; 1857, 150) and Adkins (1928, p. 126). It should be noted in this connection that *Neithea occidentalis*, as figured by Meek in the Emory's Boundary Report, is of a more trigonal outline and has straighter ribs, less divergent ventrally.

Neithea irregularis (Böse) has more prominent secondary ribs than the Arizona species. If the relation in size between the primary and minor ribs in these two species is taken into consideration, the difference is quite striking. It also appears from the illustrations of Böse (1910, Pl. 15, figs. 10–18), Kniker (1918, Pl. 2, figs. 1–6), and Adkins and Winton (1919, Pl. 11, figs. 11–15) that the primary ribs of *Neithea irregularis*, especially the second and the fifth, either remain straight to the ventral margin, or are but slightly curved, which probably justifies Kniker's placing of Böse's species in the *Neithea alpina* (d'Orbigny) group, whereas the characters of *Neithea vicinalis*, i.e. the deeper major interspaces and curved outer ribs, connect it rather strongly with *Neithea morrisi* (Pictet and Renevier).

The difference in arrangement of the tertiary ribs in their relation to the primaries in the typical forms of *Neithea irregularis* (Adkins and Winton, 1919, p. 68) and in *Neithea vicinalis* is shown in the following table.

Primary ribs	*Neithea vicinalis* Tertiary ribs	*Neithea irregularis*
First	Posterior	Posterior
Second	Anterior and posterior (weak)	Anterior
Third	Anterior and posterior (both weak)	Anterior
Fourth	Anterior and posterior (both weak)	Anterior and posterior
Fifth	Anterior (weak) and posterior	Posterior
Sixth	Anterior	Anterior

TYPE: Holotype: No. 91479.
OCCURRENCE: Perilla member, division 2b, of Lowell formation, Ninety One Hills.

Family LIMIDAE d'Orbigny

Lima Bruguière 1797.

Lima espinal Stoyanow, sp. nov.

(Plate 9, figure 8)

Holotype is an imperfect left valve with the anterior margin passing into the ventral margin as a continuous and even curve, the posterior extremity short and truncated, and the posterodorsal margin long and straight. In the middle and posterior parts of the valve the evenly rounded, broad primary ribs alternate with the much narrower and lower secondary ribs which fill the interspaces between the primaries. The most prominent or "principal" ribs, the widest and most elevated in the entire set, placed a little anterior to the center of the valve, is asymmetrically split in its ventral third on the posterior side, so that in this part of the valve it appears that two secondary ribs are present between the primaries. In the anterior fourth of the valve the ribs gradually become weak, obscure, and undifferentiated. All the ribs are ornamented with lappet-like projections arranged concentrically. It is probable that specimens with a better preserved shell-surface will show tuberculation or punctuation on the ribs.

This *Lima* essentially differs from the Cretaceous species of similar general habitus and ornamentation, described by d'Orbigny, Pictet, Woods, and Stephenson, in the nature of the uniformly wide and longer secondary ribs which fill the entire interspace between the primary ribs.

TYPE: Holotype: No. 91181.

OCCURRENCE: Espinal grit. Joserita member, division 8c, of Lowell formation. Ninety One Hills.

Lima cholla Stoyanow, sp. nov.
(Plate 9, figures 3-4)

Holotype is a left valve with the anteroventral margin partly broken. That part is shown in the illustrated paratype. Both specimens are for the greater part testiferous.

Shell thin, of rectangular-oval outline; beaks small, both ears slightly bent downward, posterior ear considerably larger; byssal notch small and shallow; anteroventral margin rounded, passing into ventral margin with a barely perceptible angulation; ventral margin a long and gentle curve; posterior extremity broad and evenly rounded; posterodorsal margin almost straight, very slightly concave; posterodorsal depression deeply excavated between the umbo and the posterior ear, shallow toward the posterior extremity; surface ornamented with radiating ribs except for the entire anterior third of the flank and the posterodorsal depression which bears only lines of growth; middle and posterior parts of the valve with 13 fine narrow ribs, separated by broad shallow interspaces which are provided with fine thread-like striae (seen only under magnification).

This species differs from the *Lima wacoensis* group in the nature of ornamentation and in the absence of the ribs in the anterior third of the shell. The European Cretaceous species that have the anteroventral region smooth, and the posterodorsal depression excavated pre-eminently near the umbones, like *Lima dupiniana* d'Orbigny and *L. subaequilateralis* d'Orbigny (1847, p. 535, Pl. 415, figs. 18-22; p. 558, Pl. 423, figs. 1-4), and *L. depressicostata* Pictet and Campiche (1858-1872, p. 158, Pl. 166, figs. 6-7) disagree with the described species in ornamentation. *L. intermedia* d'Orbigny (1847, p. 550, Pl. 421, figs. 1-5) has a similar character of ribs but it is a more densicostate species with a better rounded posterior part of the shell, a convex ventral margin, deeply excavated posterodorsal depression throughout its length, and ribs present in the anterior part of the shell. In *Lima? acutilineata texana* Stephenson (1943, p. 145, Pl. 23, figs. 1-2) the ribs are absent in the anteroventral part of the flank from the umbo to the anteroventral margin and on the posterodorsal slope.

TYPE: Holotype: No. 91491. Paratype: No. 91467.

OCCURRENCE: Cholla member, division 5e, of the Lowell formation. Ninety One Hills.

Lima muralensis Stoyanow, sp. nov.
(Plate 9, figure 2)

Holotype is the right valve of an imperfect, partly testiferous specimen of a rectangular-ovate outline, considerably inflated in the umbonal region, with posterodorsal margin convex, posterodorsal depression deeply excavated, anterior ear small, posterior ear much larger, and byssal notch strongly indicated. There are 21 ribs on the flank, straight, sharp-topped and tent-shaped in the cross-section, with broad concave interspaces adorned with fine radial striae seen only under magnification. Extreme anterodorsal region and the posterodorsal depression are ribless. Ventral part of the valve bears fine lines of growth.

This species belongs in the *Lima wacoensis* group and somewhat resembles *Lima mexicana* Böse (1910, p. 92, Pl. 14, figs. 14-15), from which it differs by its straight ribs, sharp-topped and evenly distributed throughout the valve, and a convex posterodorsal margin.

TYPE: Holotype: No. 94010.

OCCURRENCE: Basal beds of Mural limestone. Ninety One Hills.

Family ASTARTIDAE d'Orbigny
Astarte Sowerby, 1818
Astarte adkinsi Stoyanow, sp. nov.
(Plate 15, figures 12-14)

MEASUREMENTS:

	Length	Height	Thickness
Syntype No. GU27	30 mm	23 mm	16 mm.
Syntype No. GU19	29 mm	22 mm	13 mm.
Syntype No. GU32	32.5 mm	23 mm	10 mm.

This is a variable species. In the early stage of growth the syntypes No. GU19 and No. GU27 fall fairly close within the limits of *Astarte elongata* d'Orbigny, as illustrated by d'Orbigny (1847, Pl. 263, figs. 8–11), Pictet and Campiche (1866, Pl. 124, figs. 8–9), and Woods (1913, Pl. 14, figs. 2a–3), except that the beaks are placed more centrally and are relatively smaller. Very soon with growth, however, the ventral margin becomes more convex, the dorsoposterior margin more steeply inclined, and the entire shell acquires a subtrigonal outline with a lesser difference between the height and the length. In the full grown shells the dorsoposterior margin is convex and steeply inclined, the posterior extremity narrow, the ventral margin gently convex, and the anterior margin broadly rounded. The umbones are not prominent, and the dorsoposterior area between the narrow escutcheon and the flank is not distinctly separated from the latter by a very broad and obtuse angulation which extends from the umbones toward the posterior extremity. The lunule is well outlined. The greatest thickness is at the middle of the shell.

Syntype GU32 might be described as another variety except for the numerous intermediate forms through which one type merges into the other. In this shell the greatest thickness is in the dorsal half whereas the ventral half is considerably compressed. Between the thick and compressed parts, and a little below the center of the shell, there is a broad concentric depression in both valves.

A third, not illustrated modification is the one in which both the anterior and posterior extremities are relatively narrower, the shell is appreciable longer, and the umbones are more prominent.

The surface is ornamented with rather strong concentric ridges of varying elevation, and somewhat irregularly spaced, which gradually gain in width but lose in relief as they approach the ventral margin. They become very narrow toward the dorsal margin.

In the interior of the shell the anterior, ventral, and posterior margins are crenulated. The pallial line is entire, the adductor and pedal muscles are deeply impressed. In the hinge plate of the right valve the posterior cardinal tooth (5b) is very much reduced. It is attached to the nymph and followed by a trigonal oblique socket. The middle cardinal tooth (3b) is prominent, stout, vertical, and trigonal. In the upper part it bears a vertical sulcus not deep enough to make the tooth bifid. An oblique trigonal socket separates it from the massive, strong, anterior cardinal tooth (3a) which is attached to and is parallel with the lunule. This tooth is strongest in its middle part which, viewed from above, is trigonal. It is asymmetrically split by a long and shallow sulcus which is at an angle to the margin of the lunule.

TYPE: Syntypes: Nos. GU19, GU27, GU32.

OCCURRENCE: Very abundant in Perilla member, division 2b, of Lowell formation, Ninety One Hills and Guadalupe Canyon. Illustrated syntypes were collected in latter canyon.

Family CAPRINIDAE d'Orbigny

Caprina d'Orbigny, 1822

Caprina sp.

(Plate 11, figures 4–5)

The two specimens illustrated in Figures 4 and 5 of Plate 11 are internal molds collected from the thin-bedded basal layers of the Mural limestone below the massive reef with rudistids. Specimen No. 94001 (Pl. 11, fig. 4) differs from *Caprina occidentalis* Conrad (1857, p. 147, Pl. 2, figs. 1a–1c) in a greater involution and in a subcircular cross-section. It has faint longitudinal striae. Specimen No. 94002 (Pl. 11, fig. 5) is shorter and wider but of the same involution. It has a broad spindle-shaped cross-section with compressed ends. Very probably it had two bands or columns on the inner curve.

SPECIMENS: Nos. 94001, 94002.

OCCURRENCE: Basal beds of Mural limestone in hills north of Hay Flat between Bisbee and Douglas.

Family UNICARDIIDAE Fischer

Unicardium d'Orbigny, 1849

Unicardium sp.

(Plate 16, figure 4)

Right valve of a partly testiferous specimen with posterior end broken off and anterodorsal margin eroded. Ornamentation of the apical part of the valve consists of smooth concentric bands. Ventrally bands gradually gain in relief and become ridges which, however, are not strong. Ligamental groove is separated from the rest of the valve by a low ridge. Cardinal tooth is barely indicated. Elongated anterior adductor impression is limited posteriorly by a well marked deep and narrow furrow (on exposed part of the internal mold).

SPECIMEN: No. 91811.

OCCURRENCE: Collected from Quajote member, division 3b, of Lowell formation. Larger forms of *Unicardium* occur in Saavedra member, division 7i. Ninety One Hills.

AMMONOIDEA

Discussion of the genera represented in the Lower Cretaceous strata of southeastern Arizona and of the related genera is arranged in the following order.

Family PARAHOPLITIDAE Spath, 1922.

Subfamily 1. PARAHOPLITINAE Roman, 1938, emended.

Forms untuberculate in all stages of growth. Costation essentially from flexed to subradial. Suture generally cut by umbilical seam at second lateral saddle. Asymmetrical first lateral lobe.

Parahoplites Anthula, 1899, emended.
Kazanskyella gen. nov.
Sinzowiella gen. nov.

Subfamily 2. ACANTHOHOPLITINAE, subfamily nov.

Tuberculate forms. Costation essentially from radial to flexed. Suture generally cut by umbilical seam at second lateral saddle. Symmetrical first lateral lobe.

Acanthohoplites Sinzow, 1908, emended.
Immunitoceras gen. nov.
Paracanthohoplites gen. nov.
Hypacanthohoplites Spath, 1923.
Colombiceras Spath, 1923.

Subfamily 3. DESHAYESITINAE, subfamily nov.

Forms with peripheral tuberculation and ventral interruption of ribs in adolescent stage. Costation essentially sigmoidal. Suture generally more developed than in *Parahoplitinae*.

Dufrenoya Burckhardt, 1915.
Deshayesites Kazansky, 1914.

Family DESMOCERATIDAE Zittel, emend. H. Douvillé, 1916.

Beudanticeras Hitzel, 1905.

Family LYELLICERATIDAE Spath, 1931.

Stoliczkaia Neumayr, 1875.

Family PARAHOPLITIDAE Spath, 1922

Subfamily PARAHOPLITINAE Roman, 1938, emend.

According to Anthula's (1899, p. 111) original diagnosis of *Parahoplites:*

"The shell consists of appreciably inflated, about one half involute whorls with a considerably rounded venter and a moderately wide umbilicus. The whorl section is quadrate or round. The whorls bear prominent ribs some of which[27] begin at the umbilical edge as strong swellings, are

---

[27] Evidently the primary ribs.

flexed on the flank, and become stronger as they pass with a forward sweep over the venter. There is one, less often there are two, secondary ribs between every two primary ribs. They originate through intercalation between, or less frequently by branching off, the primary ribs, and are not essentially different from the latter. On the venter all the ribs are alike and uniformly spaced. . . .

"The suture, as seen in the genotype, *Parahoplites melchioris*, is composed of an external lobe, both lateral lobes, and an auxiliary lobe. All the lobes and saddles are unusually broad, stout, and rounded, with as little differentiation and branching as in *Acanthoceras*. The external lobe is somewhat more slender than the first lateral lobe, it is symmetrically divided by the siphonal saddle and usually has very short terminal branches. The first lateral lobe, which is very stout and broad, is appreciably deeper than the external lobe, it is trifid with the external branch stronger developed than the inner branch. The second lateral lobe is somewhat shorter and narrower than the first lateral lobe and appears to be asymmetric because of somewhat stronger development of its external branch. The saddles also are broad and little differentiated. The external saddle usually is as high as the first lateral saddle, and is divided by a shallow lobule into two unequal branches which are comparatively little differentiated. A similar development is seen in the first lateral saddle which is at the umbilical edge and, with the second lateral saddle, outlines the second lateral lobe.[28]" (Translated).

Without comments, Anthula separated the numerous species, all of which he assigned to *Parahoplites*, into two groups: 1. The group of *Parahoplites melchioris*, and 2. The group of *Parahoplites aschiltaensis*.

Jacob (1905, p. 407) placed the entire group of *Parahoplites aschiltaensis* of Anthula in "*Douvilleiceras*," i.e., partly in *Cheloniceras* of the present literature (with *Ammonites martini* d'Orbigny, *A. cornuelianus* d'Orbigny, *A. nodosocostatus* d'Orbigny, and *A. mammillatus* Schlotheim).

Sinzow (1908) restricted *Parahoplites* to those species that have a smooth shell in the young stage and later develop untuberculated ribs, in the suture of which the first lateral lobe is tripartite and deeper than the external lobe (this according to Anthula's original diagnosis). He emphasized the asymmetrical nature of the sutural branches of *Parahoplites* and compared its suture to that of "*Leopoldia*." For Anthula's group of *Parahoplites aschiltaensis* Sinzow established a separate genus *Acanthohoplites*.[29]

Subsequently, Kazansky (1914, p. 66–77) commented with much acumen on the relationship between *Parahoplites* and *Acanthohoplites*:

"Speaking generally, *Parahoplites melchioris* is a form that stands apart from all other species which were attached to *Parahoplites* by Anthula. Characteristically simple and broad lobes and saddles, a forward sweep of ribs on the venter, and reduced ornamentation on the young whorls, very definitely separate *Parahoplites melchioris* from all other species placed by Anthula in the group of that name.

"If the characteristics of Anthula's genotype alone are accepted as diagnostic for the genus, the species referred by Sinzow to *Acanthohoplites* will appear sufficiently separated from *Parahoplites melchioris* to justify an independent genus. If, on the other hand, *Parahoplites* is understood to embrace the entire *Parahoplites melchioris* group in Anthula's interpretation, Sinzow's genus completely blends with *Parahoplites*. The diagnostic characteristics of *Acanthohoplites*, as outlined by Sinzow, are essentially in the tuberculated nature of the young whorls and a lesser assymmetry of the first lateral lobe as compared with *Parahoplites melchioris* the first lateral lobe of which not always is very asymmetric yet is wider and blunter than in *Acanthohoplites*. If the interpretation

---

[28] In the text: ". . . und mit dem auxiliare den Nahtlobus bildet." An evident discrepancy, not in agreement with the preceding discussion.

[29] Regarding the spelling "*Acanthohoplites*" versus "*Acanthoplites*," now nearly in general use (Kilian, 1913, p. 345–346; 1915, p. 43; Spath, 1923, p. 4; Renngarten, 1926, p. 25; Roman, 1938, p. 348), I am inclined to believe that the ungrammatical "*Acanthoplites*"—"emended" by Kilian, generally a poor speller (Sintzow for Sinzow; *Dufrenoyia* and *Dufrenoya* in the same paper, etc.,—is a *lapsus calami* and should be attributed rather to editorial vicissitudes, which probably were influenced by Hyatt's equally ungrammatical "*Metacanthoplites*" in Zittel's textbook (1902 edition), than to the approval of Sinzow, a man of solid classical education, who plainly stated in his original definition (Sinzow, 1908, p. 457–458): "Die Arten aber, die Anthula zu *Parahoplites* rechnet, bei welchen eine höckerige Sculptur beobachtet wird, die derjenigen der Vertreter des *Douvileiceras* ähnlich ist, scheide ich zu die Gattung *Acanthohoplites* aus." Evidence of the derivation of the word is quite clear. Etymologically "*Acanthoplites*" is more incorrect than "*Acanthclymenia*" or "*Acanthceras*" would be for *Acanthoclymenia* and *Acanthoceras*, since neither of the two parts of the combined word can claim "h" as its own. Assuming that "th" stands for the Greek theta of the correctly latinized first part of the word, *oplites* is then an incorrect transliteration of the Greek '*oplites* which should be transliterated as *hoplites*, and if the latter is the case, then the theta is represented only by "t." (Compare Spath, 1939a, p.238, footnote).

of *Parahoplites melchioris* is thus restricted, almost all the species of Anthula's group *"melchioris"* are to be placed with *Acanthohoplites*." (Translated).

COSTATION OF PARAHOPLITINAE. Costation of the subfamily Parahoplitinae, as interpreted in this paper, requires some comment. Within the family Parahoplitidae, and often during the individual development, the costation varies through all gradations from sigmoidal, generally with a forward convexity in the umbilical half of the rib and with a forward sinuosity in the ventral half of it, as in *Deshayesites deshayesi* (d'Orbigny, 1840, Pl. 85, fig. 1), to almost perfectly radial ("rectiradiate" of Spath). Between these two extreme types, as one of the possible modifications of sigmoidal ribs, are the ribs with two sinuosities. A rib of this nature will have consecutively sinuosity-convexity-sinuosity between the umbilical edge and the peripheral margin. Here the degree of flexuosity and position of convexity should be noted, as the latter may occur below, at, or even extend above (as in the "rursiradiate" *Parahoplites treffryanus*," Anthula, 1899, Pl. 8, fig. 6a, non *Am. treffryaanus* Karsten), the midflank of the whorl.

If, as an example, the ribs of *Cleoniceras cleon* (*Ammonites bicurvatus* d'Orbigny, 1840, Pl. 84, fig. 3) are "flexiradiate", according to Spath's (1923, p. 8) terminology, and the ribs of *Deshayesites grandis* Spath (1930, p. 428, Pl. 17, fig. 2a) are "falcoid", the difference between the two is obviously only in the degree of flexuosity, that is, the upper sinuosity is greatly reduced in the later stages of *Deshayesites grandis* Spath.

In the ammonites with flexed costation, the intervening convexity, or the bulge, of the ribs may be very short, as in *Leopoldia castellanensis* (d'Orbigny, 1840, Pl. 25, fig. 3), for instance, or it may be very long, as in *Neoharpoceras hugardianum* (d'Orbigny, 1940, Pl. 86, fig. 1).

Especially characteristic for certain forms from the Aptian of the Transcaspian Region and the Caucasus of Russia, and from Arizona, which are more or less related to the true *Parahoplites*, are comparatively narrow ribs with a long and low convexity over the greater portion of the midflank, and with very short, shallow sinuosities restricted to the umbilical and peripheral parts of the flank. In *Parahoplites melchioris* Anthula, (1899, Pl. 8, figs. 4a and 5a) the ribs of this nature are barely indicated, on the other hand, in the forms from Mangyshlak and the Caucasus placed by Sinzow in *Parahoplites*, like *Parahoplites "melchioris"* Sinzow (1908, Pl. 2, figs. 1-4) and *Parahoplites multicostatus* Sinzow (1908, Pl. 2, figs. 5, 7, 11) such ribs are appreciably distinct from the more flexed ribs of Anthula's genotype.

Further, in certain Parahoplitinae the ribs of this kind tend toward radiality through the reduction and almost total disappearance of the median convexity, but since the marginal sinuosities are never completely eliminated, the ribs eventually become subradial and prorsiradiate. Approximation to this extreme type of costation is indicated on the unchambered part of the last whorl of *Parahoplites "melchioris"* Sinzow (1908, Pl. 2, fig. 1). In *Kazanskyella*, gen. nov., (Pl. 17, figs. 2-8) the sinuosity of the ribs is barely perceptible in the adult whorls. The costation of this nature is referred to in this paper as "parahoplitan," as distinguished from the essentially sigmoidal or sickle-shaped "deshayesitan" ribbing.

The mode of distribution of the ribs on the earlier, penultimate, and ultimate whorls of Parahoplitinae is alone characteristic enough to justify the establishment of a separate subfamily. It should be noted that three regions are clearly distinguishable in the costation of *Parahoplites melchioris* Anthula (1899, Pl. 8, figs. 4a and 5a): (1) the region of earlier whorls with simple, sparsely set ribs; (2) the region of "crowding" which extends from the beginning to the middle of the last whorl and within which the ribs, partly flexed or even subsigmoidal, are noticeably crowded, the secondaries appear in twos, and their origin is indefinite (*i.e.*, branching or intercalating); (3) the region corresponding with the later part of the last whorl in which the ribs again become sparsely set and regularly spaced. The ribs are less flexed here, and there is only one secondary rib between every two primary ribs.

In *Kazanskyella* the region of "crowding" occurs at a very early stage, at diameters of 12-16 mm., and occupies a rather limited portion of the entire shell. The last region of the sparsely set and regularly alternating primary and secondary ribs is already established above 22-24 mm. diameter.

In the earlier whorls of *Sinzowiella*, gen. nov., (Pl. 18, figs. 1-17) the ribs are most densely set, from there on, the interspaces between the ribs increase in width very gradually. Consequently, there is no sharply differentiated region of costation as far as the distribution of ribs is considered, nevertheless, a region analogous to "crowding" exists in the part of the shell in which the simple ribbing is suddenly substituted by branching ribs and in turn gives way to a more regular costation of the primary ribs which uniformly alternate with single intercalating secondary ribs, as in *Parahoplites* and *Kazanskyella*.

The presence of a region of rather crowded and branching ribs which separates the earlier whorls with a simple ornamentation of uniform ribs from another but later region of the mature shell with simply alternating primary and secondary ribs, is regarded here as one of the most characteristic features in the costation of Parahoplitinae, a feature not apparent either in Deshayesitinae or Acanthohoplitinae. It should also be noted that in Parahoplitinae the maximal flexuosity of the ribs is attained in the early stages of growth, whereas in Acanthohoplitinae the early radial costation changes to the flexed ribbing in the mature whorls.

SUTURE OF PARAHOPLITINAE: There is a considerable discrepancy between Anthula's diagnosis of the suture of *Parahoplites melchioris* and its rather schematic and probably not very accurate drawing (Anthula, 1899, Pl. 8, fig. 4c; Roman, 1938, p. 350, fig. 325). It also is uncertain whether the illustrated suture was taken from a single specimen and was not too corroded for a correct interpretation. This is discussed further under *Parahoplites*. Some facts that are inferred from the perusal of Anthula's paper require attention at this time. There is a definite statement about the trifid nature of the first lateral lobe and its somewhat asymmetric outline. This indeed is supported by the illustration, and it has been generally accepted by the writers who had to interpret Anthula's genus that this asymmetry is not very great. An additional corroborant may be derived from the frequent allusion in Anthula's very brief description of the suture of the other species of his *"Parahoplites,"* now generally placed under *Acanthohoplites*, to the effect that these sutures are not essentially different from the suture of his genotype. This assertion may be significant since nearly all Caucasian and Transcaspian species of *Acanthohoplites* described later by Sinzow (1908) have a more or less symmetrical first lateral lobe with pointed and well-drawn lobules. There seems to be sufficient direct and circumstantial evidence that the first lateral lobe of *Parahoplites melchioris* Anthula is asymmetrically trifid.

No less significant is Sinzow's statement in the very beginning of his treatise on *Parahoplites* and *Acanthohoplites* (Sinzow, 1908) that the first lateral lobe of *Parahoplites* is similar to that in *Leopoldia* (d'Orbigny, 1840, Pl. 22, fig. 3; *compare* Sarasin, 1897, p. 775, fig. 8). The entire suture of *Leopoldia* has very little in common with the suture of *Parahoplites*. The explanation is that Sinzow did not fail to notice the essential difference between the suture of *Parahoplites*, as described and figured by Anthula, and the suture of certain ammonites which he placed under *Parahoplites* in his monograph. For instance, unlike the first lateral lobe with nearly parallel sides and very short terminal branches of Anthula's genotype, in the suture of *Parahoplites "melchioris"* Sinzow (1908, Pl. 2, figs. 1, 2, 4; *see also* Pl. 18, figs. 9 and 10 of this paper) the posterior half of the first lateral lobe is asymmetrically separated into three long and slender strongly developed branches. It is noteworthy that he compared the suture of *Parahoplites maximus* Sinzow (1908, p. 465, Pl. 1, fig. 3) with that of *Leopoldia leopoldi* and not of Anthula's genotype.[30]

Analysis of the material from southeastern Arizona and the study of literature leads to the conclusion that there are two types of suture in Parahoplitinae. The one with a longer and trifid first lateral lobe, the asymmetry of which is indicated rather by the uneven development of the branches and lobules in its terminal part than by the asymmetrical nature of the sides and the attitude of the lobe itself, and the other with a broad, short, and shallow first lateral lobe which becomes greatly asymmetrical in the adult stage. Since these types of suture seem to be sufficiently persistent and develop independently, I distinguish in the sutures of the following genera: 1. *Parahoplites* Anthula (external lobe with parallel sides; first lateral lobe subsymmetrical in attitude, broad and with subparallel sides anteriorly, tending toward tripartition with asymmetrically developed branches in the posterior part; bifid symmetry of saddles not strongly indicated). 2. *Sinzowiella*, gen. nov. (external lobe with diverging sides; first lateral lobe slanting, strongly trifid in the young stage, but broad, short and asymmetrical, with tripartition considerably obliterated, in the adult whorls; saddles

---

[30] Less fortunate is the comparison with the more symmetrically developed suture of *Ammonites campichei* Pictet and Renevier (1858, p. 25, Pl. 2, fig. 2).

symmetrically bifid). 3. *Kazanskyella*, gen. nov. (external lobe broad and diverging; first lateral lobe broad, asymmetrical, strongly slanting away from the external saddle, with tripartition lost in the earliest stages of growth; saddles symmetrically bifid).

As examples of the first type, the sutures of *Parahoplites melchioris* Anthula (1899, Pl. 8, fig. 4c) and *Parahoplites inconstans* Riedel (1937, Pl. 14, fig. 13) may be mentioned. The second type of suture is best exemplified in *Kazanskyella* and *Sinzowiella*.

Variations in the first lateral lobe of *Parahoplites* are well seen in the parahoplitan material described by Sinzow. Among the illustrated specimens are forms both with the slender and with the broad first lateral lobe. It is noteworthy, however, that their essentially trifid character either persists or at least is recognizable from the young to the adult stages (Sinzow, 1908, Pl. 1, figs. 1, 3, 4, 7, 10, 11, 13; Pl. 2, figs. 1, 4, 11). In *Kazanskyella* and *Sinzowiella* the slender trifid nature of the first lateral lobe is lost either in the youth or in the early maturity respectively.

Parahoplites Anthula, 1899, emend.

GENOTYPE: *Parahoplites melchioris* ANTHULA (1899, p. 112, Pl. 8, figs. 4a–5b).

The genus *Parahoplites* has often been used as a "catch all" group for a rather large series of species, many of which probably are not congeneric with Anthula's genotype. The diagnosis of *Parahoplites* is essentially as follows: Shell smooth in the early stage. In the immature stage ribs are simple, either not differentiated or with incipient differentiation into primary and secondary ribs. The early mature stage is marked by crowding of flexuous ribs, which results from a rapid appearance of secondary ribs that originate through branching or intercalation, or both. In this stage there may be more than one secondary rib between any two primary ribs. In adult stage, the single secondary ribs alternate with the primary ribs, and all the ribs retain flexuousity to an appreciable degree. At a certain stage the ribs develop a forward sweep on the venter, they may be thicker in the ventro-lateral part and stronger on the venter. At no stage are the ribs depressed or interrupted on the venter. Likewise, at no stage is there any tuberculation. The whorl section is never polygonal or strongly angular. In the genotype the greatest thickness of the whorl is below its midheight.

It has been pointed out already that Anthula's original drawings of the suture of the Caucasian Cretaceous ammonites, assigned by him to *Parahoplites*, are in the main too diagrammatic and not sufficiently accurate. There can hardly be doubt, however, that one of the outstanding characteristics of the "reduced" parahoplitan suture is the first lateral saddle, which is remarkably imitative of, and but little smaller than, the external saddle. These are the only two complete saddles on the flank. The great similarity between them is not quite apparent in the original sketches of Anthula (1899, Pl. 8, fig. 4c) and in the later and better executed reproduction of Roman (1938, p. 350, fig. 325), probably because Anthula's original specimen was somewhat worn, a condition indicated by the discrepancy between the right and left halves of the suture illustrated by Anthula and Roman. Much better preserved sutures are seen in Sinzow's photographic pictures of the forms more or less related to *Parahoplites melchioris* Anthula. Both the pictures of Sinzow and the right half of the suture as figured by Roman show the first lateral saddle formed similarly to the external saddle, in agreement with Anthula's text, but not tripartite as drawn in his figure.

The following brief revised outline of the suture of *Parahoplites melchioris* is based on the illustrations given by Anthula and Roman: The little divided external lobe has parallel sides. The first lateral lobe, with subparallel sides, has three uneven terminal branches which make it asymmetrical, although the degree of asymmetry has been variously interpreted by different writers. The external and first lateral saddles are similar in composition and asymmetrically bipartite (as in the right half of Roman's illustration). On the flank the suture terminates in the asymmetrical and oblique second lateral lobe with a portion of the second lateral saddle.

Parahoplites Anthula differs from *Kazanskyella* and *Sinzowiella* in the distribution and nature of the ribs, and the conspicuously trifid asymmetrical first lateral lobe.

Kazanskyella Stoyanow, gen. nov.

GENOTYPE: *Kazanskyella arizonica* STOYANOW, sp. nov.

In 1914 Kazansky described and illustrated a few conspecific forms from the Aptian of Daghestan, Caucasus, which he identified with *Parahoplites melchioris* Anthula. His brief description (Kazansky, 1914, p. 90-91, Pl. 5, figs. 76-78) deals only with the immature representatives of this assemblage:

"These specimens are distinguished by a nearly perfect circular whorl section and almost smooth whorls. The typical ornamentation begins to appear at a diameter of 6-10 mm. and consists of low, round, and distinctly flexed ribs already differentiated into the primary and secondary ribs not interrupted on the venter. There is no evidence of tuberculation in these young specimens." (Translated).

Kazansky rendered no description of mature specimens, some of which attained 76 mm. diameter, nor did he describe the suture of these ammonites from Daghestan. It appears, however, that the examples figured by Kazansky in three successive stages of growth essentially disagree in the nature of ornamentation with *Parahoplites melchioris* Anthula and with *Parahoplites* in general, as defined in this paper.

The genus *Kazanskyella* is established for the above listed forms of Kazansky and for the congeneric species from the Lowell formation of southeastern Arizona. As pointed out by Kazansky, the earliest whorls are smooth in these forms. Next follow the whorls with ribs early differentiated into the primaries and secondaries. As already mentioned, the region of denser costation with somewhat flexed ribs, in which the primaries bifurcate and the closely set secondaries appear either as single ribs or in twos between the primary ribs, is very limited, being restricted to the shell between the 14 and 16 mm. diameters. With growth the ribs rapidly become subradial on the flank, and are slightly bent forward only at the peripheral margin. On the venter, the ribs do not form a strong forward sinus so characteristic of *Parahoplites melchioris* Anthula (1899, Pl. 8, figs. 4b, 5b), but instead arch gently and broadly forward between the peripheral margins. In the adult specimens the greatest thickness of the last whorl is about the midflank and not near the umbilical edge as in Anthula's genotype, and the subradial ribs are coarser and more distantly spaced. The suture is cut by the umbilical seam at the second lateral saddle. It has a strongly asymmetrical, broad and oblique, first lateral lobe that is not distinctly tripartite even in the earlier whorls, whereas in the genotype of *Parahoplites* the first lateral lobe is not conspicuously asymmetrical (Roman, 1938, p. 350, fig. 325), and in the specimens from the Caucasus and Mangyshlak identified with *Parahoplites melchioris* Anthula by Sinzow (1908, Pl. 2, figs. 1-4; see also Pl. 17, figs. 9 and 10 of this paper), it is asymmetrically trifid. The external and first lateral saddles are bifid.

At present two species are included in this genus: *Kazanskyella arizonica*, sp. nov., from southeastern Arizona, and *Kazanskyella daghestanica*, nom. nov. = *Parahoplites melchioris* Kazansky (non Anthula) from Daghestan, Caucasus.

Kazanskyella arizonica Stoyanow, sp. nov.

(Plate 17, figures 2-8)

MEASUREMENTS.

	Syntype No. 91749	Syntype No. 91111	Syntype No. 91748	Syntype No. 91747
Diameter	17	22	64	66
Greater radius	11	14	40	41
Lesser radius	7	8	24	25
Height of last whorl	9	11	24	25
Thickness of last whorl	8	10	24	25
Width of umbilicus	4	5	18	19

Shell of this species is a little less than one half involute, with a deep umbilicus and steep umbilical wall especially so in the younger stages of growth. Whorl section, round in the young whorls, passes to oval as the slightly inflated flanks become flatter with growth. With the rapid increase of the ribs in size the mature whorls attain an ovate section. Peripheral area is broadly arched.

Costation is clearly indicated at 6 mm. diameter, where it consists of uniformly straight coarse ribs slightly inclined forward at the peripheral margin. The region of somewhat flexed and "crowded" ribs, which results from the sudden introduction of one or two closely set secondary ribs between every two primaries, is limited to the shell below 16 mm. diameter. About and above this diameter all the ribs continue low on the flank to the umbilical edge, but the primaries are somewhat stronger and extend a little lower. At this stage all ribs are straight in the umbilical half of the flank, but above the midflank they bend slightly forward and form a shallow sinuosity. All ribs thicken very gradually toward the peripheral margin and are thickest on the peripheral area. Beginning with diameter of 20 mm. the ribs are straight and more attenuated on the flank, although a somewhat flexed rib occurs occasionally. Between diameters of 30 and 45 mm. the primary ribs, radial and distantly spaced on the flank, appear low on the umbilical wall. Above 45 mm. diameter they form a single, barely perceptible, shallow curve between the umbilical edge and the peripheral margin. The secondary ribs appear higher on the flank with age. Above a diameter of 30–32 mm. they are very thin at or below the midflank but rapidly increase in thickness toward peripheral margin. On the peripheral area of the unchambered part of the last whorl the ribs are broad, steep in front and gently sloping backward.

In the suture, as observed between diameters of 13 and 24 mm., the short external lobe rapidly widens anteriorly. The broad external saddle, symmetrically separated into two short bipartite branches in its anterior part, has asymmetrical sides, the outer side is short whereas the inner side is long and slopes obliquely inward. The outer side of the broad and asymmetrical first lateral lobe is not sharply differentiated from the inner side of the external saddle, whereas its inner side rises abruptly to form the outer side of the first lateral saddle. This asymmetry obscures the trifid parahoplitan nature of the first lateral lobe and is further augmented by its slanting terminal part with the narrow rounded lobules which gradually increase in depth inward, and of which the last and innermost lobule is slightly bifid. The first lateral saddle is symmetrically bifid and, like the external saddle, is separated into two short and stout bipartite branches. Though nearly as broad as the latter saddle, it is shorter and has subparallel sides. The second lateral lobe is small, narrow, rounded, little differentiated, and inclined outward. The umbilical seam cuts the suture at the small second lateral saddle which apparently is asymmetrically bipartite. In the observable earlier sutures of the inner whorls the terminal lobules of the less asymmetrical first lateral lobe are subequal and relatively short. In the last whorl, as is seen in the idiotype, No. 91241, of 92 mm. diameter, with the unchambered part preserved, the more differentiated suture is essentially as described.

Kazanskyella arizonica differs from *Parahoplites melchioris* Anthula in the whorl section which, because of the greatest whorl thickness at the midflank, is symmetrically oval. Its ornamentation differs in the predominantly straight ribs on the flank, lack of a forward sinus in the ribs on the peripheral area, and reduction of the region of "crowded" ribs which, besides, is restricted to smaller diameters. The most distinguishing difference in the suture of these two ammonites is in the nature of the first lateral lobe.

The younger whorls of *K. arizonica* and *K. daghestanica* seem to be much alike. The primary ribs of the second species acquire the radial character earlier on the last whorl, and the latter is more inflated. The suture of *K. daghestanica* has not been described sufficiently to make an adequate comparison.

Type: Syntypes: Nos. 91749, 91111, 91748, 91747. Idiotype (not illustrated) No. 91241.

Occurrence: Syntypes collected from Cienda limestone, Pacheta member, division 9c, of Lowell formation, Ninety One Hills. Idiotype came from the same stratum at the top of Black Knob Hill.

Sinzowiella Stoyanow, gen. nov.

Genotype: *Sinzowiella spathi* Stoyanow, sp. nov.

The shell of this genus, inflated and with a nearly round whorl section in the young stages, gradually develops, together with the increase in involution and in whorl height, flat flanks which in the early mature whorls of the genotype converge toward the arched venter. With the further growth, the flanks gently curve, the greatest thickness of the whorl migrates toward the midflank, and the whorl section becomes suboval.

The smooth inner whorls early acquire a fine densicostate ornamentation of simple ribs slightly inclined forward on the flank and broadly arching on the periphery. The region of flexed and bifurcating single ribs precedes a little the introduction of the secondary ribs and is limited to the early adolescence. The costation of mature whorls consists of slightly flexed unbranching primary ribs which alternate with the single secondaries.

The suture of the inner whorls has a short external lobe with diverging sides, symmetrically bifid external and first lateral saddles, and an asymmetrically trifid first lateral lobe with parallel sides which is strongly inclined inward. With growth the inner and the outer lobules of the first lateral lobe become shallowly bifid. With the gradual shortening of the odd middle lobule in the mature whorls, the entire first lateral lobe acquires a broad and asymmetrically fan-shaped appearance. The second lateral lobe is small. The suture is cut by the umbilical seam at the second lateral saddle.

The development of a shell with flat flanks, the fine densicostate ornamentation in the early whorls, and the rapid change of the first lateral lobe from trifid to asymmetrically broad, serve to distinguish this genus from *Parahoplites*, as defined in this paper, and from *Kazanskyella*.

Sinzowiella spathi Stoyanow, sp. nov.

(Plate 18, figures 1–17)

MEASUREMENTS:

	Syntype No. 91203	Syntype No. 19083	Syntype No. 91099	Syntype No. 91074	Syntype No. 91112
Diameter	8.5	9	12	16	21
Greater radius	5	5	7	10	13
Lesser radius	3.5	4	5	6	8
Height of last whorl	4	5	5.5	7.8	10.5
Thickness of last whorl	3.5	4.5	5	6.8	
Width of umbilicus	2.5	2.8	3	4.5	

The shell is nearly evolute, smooth, and with a round whorl section in the first three volutions. Below a diameter of 10 mm. it is about one fourth involute, has a deep and narrow umbilicus and inflated flanks which are undifferentiated from the broadly arched peripheral area and from the steeply inclined but curved umbilical wall.

The costation is introduced at a diameter of approximately 5–6 mm. At this stage the densely spaced and nearly straight fine single ribs are weak below the midflank but gain in strength toward the venter, which they cross in broad and low arches. However, a forward curve above the midflank and a broad forward peripheral sweep are acquired very early, and at a diameter of about 10 mm., a not sharply defined forward sinus is formed on the venter. At and above this diameter, with the flattening of the flanks and the increase in involution and in whorl height, the ribs become more flexed on the flank, appear in the umbilical wall, form an appreciable swelling above the umbilical edge, and bifurcate low on the flank. Next appear the secondary ribs, one between every two branching primaries. In this way the region of "crowded" costation is formed. At a diameter of about 20 mm. the branching of the primary ribs ceases. The primary ribs, no longer inflated above the umbilical edge, develop a low forward convexity below the midflank and a shallow forward sinuosity above it. A costation of unbranching slightly flexed and subradial primary ribs, with the single and slightly bent forward secondary ribs between every two primaries, is established. The primaries increase very little in thickness between the umbilical edge and the peripheral margin. The secondary ribs are very thin below the midflank but equal the primaries in thickness at the peripheral margin and on the venter. In the largest observed whorls the primary ribs form a low middle convexity and sinuosities above the umbilical edge and below the peripheral margin on the gently curved flank, and all the ribs are broadly arched on the venter.

In the suture of young whorls the short external lobe has two branches in each of its diverging sides. Both sides of the external saddle slant away from its broadly rounded anterior part divided by a short lobule into short and stout symmetrically bifid halves. The oblique first lateral lobe, subsymmetrically trifid in its terminal part, has a small lobule on the outer of its subparallel sides. The first lateral saddle, with subparallel sides, is more symmetrical than the external saddle and likewise is divided into wide and short bifid halves. The second lateral lobe is small, narrow, with an incipient tripartition, and nearly concentric with the umbilical edge. The second lateral saddle is short, flat and slightly bifid anteriorly, its outer side is nearly vertical wheras the inner side is a curve cut by the umbilical seam. As has been mentioned in the diagnosis of the genus, the inner and the outer lobules of the first lateral lobe become slightly bifid with growth. In the further development its inner side becomes short and the outer side not sharply differentiated from the slanting inner side of the external saddle. As a result, the first lateral lobe is asymmetrically broad and denticulated in the mature whorls.

The ribbing in the younger whorls of *Sinzowiella spathi* has a certain resemblance to that in the form from Mangyshlak placed by Sinzow (1908, p. 474, Pl. 3, figs. 16–17 only) in his species *Sonneratia rossica*. This form differs from the described species in a greater sinuosity of the ribs in the upper part of the flank and an oval whorl outline at the diameter of 36 mm. Although the suture of Sinzow's specimen is unknown, most probably it is not congeneric with the large and septate form of *Sonneratia rossica* figured on the same plate. A form, also from Mangyshlak, described by Sinzow (1908, p. 464, Pl. 1, fig. 3, only) as *Parahoplites maximus* has densicostate inner whorls, and its convergent flanks and the uniformly thick, subradial and flexed, ribs in the mature whorl are as in *Sinzowiella*. Since, however, as already discussed under the subfamily Parahoplitinae, the first lateral lobe of this form has a nearly symmetrical attitude and a strong tripartition at full maturity, it should be retained in *Parahoplites*.

TYPE: Syntypes: Nos. 91203, 91083, 91099, 91074, 91112, 91113, 91087, 91210, 91110, 91095, 91100.

OCCURRENCE: All syntypes collected from Cienda limestone, Pacheta member, division 9c, of Lowell formation, Ninety One Hills, where this species is found with *Kazanskyella arizonica*.

Sinzowiella ? sp.
(Plate 23, figs. 7–13)

In the Tusonimo limestone, a few feet below the Cienda limestone with *Sinzowiella spathi*, occur numerous but poorly preserved and fragmentary specimens which are tentatively placed with *Sinzowiella*.

In the earlier stages, these forms have a narrow whorl and very flat flanks which, with growth, converge toward the venter through an increase of the whorl in thickness below the midflank. A few representative examples are illustrated on Plate 23. (1) Two conspecific specimens Nos. 91092 and 91209 (Pl. 23, figs. 13 and 12) of the whorl height 13–16 and 38–44 mm. respectively, have the secondary ribs low on the flank. In the larger of these two fragments the flanks gradually change from flat to gently curved within the indicated whorl heights. (2) Specimen No. 91202 in which the younger whorls (not illustrated) have parallel flanks and ribs with a forward sweep on the venter, but at the whorl height of 26–28 mm. develop convergent flanks and nearly straight ribs on the broad and arched peripheral area (Pl. 23, figs. 7–9). (3) A variety in which the ribs with a forward bend on the flat flank are substituted by subradial ribs at a much later age than in *Sinzowiella spathi*. Of the three collected specimens of this form the somewhat weathered and slightly deformed example, with a shallow trifid first lateral lobe at the whorl height of 14 mm., is shown in Figures 10–11 of Plate 23. The ribbing of this form resembles that in the specimen from Mangyshlak identified by Sinzow (1908, Pl. 1, fig. 11, only) as *Parahoplites grossouvrei* Jacob, in which, however, the suture has a slender first lateral lobe with long asymmetrical branches, as in the other species placed by Sinzow in *Parahoplites*.

SPECIMENS: Nos. 91202, 91209, 91092, 91126.

OCCURRENCE: Tusonimo limestone, Pacheta member, division 9d, of Lowell formation, Ninety One Hills.

Subfamily ACANTHOHOPLITINAE Stoyanow, subfamily nov.

In an essay on the ammonite fauna and stratigraphy of Clansayes, France, Jacob (1905, p. 407) placed under *Douvilleiceras* Grossouvre a number of species originally included by Anthula (1899, p. 110) in his group "*Parahoplites aschiltaensis*" and at the present time generally interpreted as members of Sinzow's genus *Acanthohoplites*. Among the species assigned to *Douvilleiceras*, which in his opinion belonged in Anthula's group, Jacob included *Ammonites martini* d'Orbigny, *A. cornuelianus* d'Orbigny, *A. nodosocostatus* d'Orbigny, and *A. mammillatus* Schlotheim which, except for the last species, all belong in Hyatt's genus *Cheloniceras*.

In view of the fact that in the early stage of growth *Acanthohoplites* is similar to the adolescent or even mature shell of *Cheloniceras*, and because there is some discrepancy in the current literature in connection with the evaluation of certain characteristics common to both genera, a brief discussion of Spath's family Cheloniceratidae (Spath, 1923, p. 64) is given here.

The establishment of this family apparently was based upon the selection of *Ammonites royerianus* d'Orbigny (1840, p. 365, Pl. 112, figs. 3, 4, 5) as the type of *Cheloniceras* according to a footnote written by Stanton in his commentaries on Hyatt's incompleted manuscript of the *"Pseudoceratites of the Cretaceous"* interrupted by his death (Hyatt, 1903, p. 101). The footnote in question reads as follows:

"In the manuscript a sheet is inserted just before *Vascoceras* with the heading 'Cosmoceratida,' followed by 'In family description notice resemblance of form to Aspidoc. of Jura as more remote than to *Cheloniceras* of the Cretacic.' Another memorandum bears pencil-sketch copies of d'Orbigny's figures of *Ammonites royerianus* (Pal. Fr. Terr. Crét., 1, pl. 112, figs. 3, 4) labeled *Cheloniceras royerianus*, indicating that he had probably selected this species as the type of a new genus...."

In 1915 Nikchitch, then engaged in studies of the Cheloniceratidae from the Aptian strata on the northern side of the Caucasian Mountains, arrived at the conclusion, based on the examination of early whorls of many forms of the group of *Cheloniceras cornuelianum* (d'Orbigny), that *Cheloniceras royerianum*, as described and illustrated by d'Orbigny, is *C. cornuelianum* in its early stage of development. Commenting on the inner whorls of Caucasian examples of *C. cornuelianum* at diameters between 5 and 10 mm., he added:

"All the characteristics of this stage precisely correspond with those of *Ammonites royerianus* d'Orbigny and I believe that *A. royerianus* is not an independent species but represents a young specimen of *Douvilleiceras cornueli* (d'Orbigny).[31] My opinion is further supported by the fact that this little specimen described by d'Orbigny does not show the body chamber, and, so far, is known only in one specimen." (Nikchitch, 1915, p. 3, 4, 13, 50) (Translated).

It may be mentioned in this connection that the types of *A. cornuelianus* and *A. royerianus*, although contributed by two different collectors, both came from the same general area and the same stratigraphic horizon of the Cretaceous of France. According to d'Orbigny's original description of *A. cornuelianus*: "Cette charmante espèce a été decouverte par M. Cornuel, à Louvemont, près de Wassy (Haute-Marne), dans l'argile à plicatules, terrain néocomien le plus supérieur. Je l'ai également de Saint-Paul-Troix-Châteaux de la même couche," and regarding *A. royerianus*: "M. Royer, à qui je dois la communication de cette Ammonite, l'a trouvée à Bailly-aux-Forges, arrondissement de Wassy (Haute-Marne), dans l'argile à plicatules qui forme la couche la plus supérieur de l'étage néocomien" (d'Orbigny, 1840, p. 365 and 366).

D'Orbigny's assertion that his illustration of *A. royerianus* represents this ammonite in natural size is incorrect. Figures 3 and 4 of Plate 112 (d'Orbigny, 1840) instead of being 12 mm. in diameter, as stated in the text, are of 20 mm. diameter as measured by calipers. If considered to be of 12 mm. diameter and reduced to that size, d'Orbigny's figures 3 and 4 will exactly correspond with the figures 1a–1c in the Plate 1 of Nikchitch's memoir (1915), which illustrate young specimens (diameter 7–12 mm.) of *Cheloniceras cornuelianum*.

I do not question the correctness of Nikchitch's interpretation. Although the relation of these two ammonites was not discussed by d'Orbigny, it may not be quite coincidental that he described in an immediate succession, and figured on the same plate, two forms that had come from two different private collections.

Fortunately enough, *A. cornuelianus*, well known in the adult stage and often quoted in a number of publications, was described first of the two forms and therefore has priority in the name over *A. royerianus*.

---

[31] Since the original name of this species is *"cornuelianus,"* when attached to the generic *"Douvilleiceras"* or *"Cheloniceras"* it should be *"cornuelianum"* and not *"cornueli"* as has been used by some writers.

Nikchitch separated the cheloniceratids of the Caucasian Aptian into two groups:

A. Group of *Cheloniceras cornuelianum*. The species included in this group, *C. cornuelianum* (d'Orbigny) and *C. seminodosum* (Sinzow), and their varieties, possess the following characteristics:
   (1) Primary ribs bear two conical tubercles.
   (2) Primary ribs branch at the superior, and at larger diameters sometimes at the inferior tubercles.
   (3) Anterior branches of the primary ribs are either stronger or as strong as the posterior branches.
   (4) Often there is a constriction anterior to a primary rib or to its anterior branch.

B. Group of *Cheloniceras martini* and *Cheloniceras tschernyschewi*. In this group are included: *C. martini* (d'Orbigny), *C. tschernyschewi* (Sinzow), *C. subnodosocostatum* (Sinzow), and *C. buxtorfi* (Jacob). The distinguishing characteristics of this group are:
   (1) Primary ribs bear three tubercles.
   (2) Primary ribs branch only at the upper tubercles.
   (3) Anterior branches of the primary ribs are either less strong or equal in strength to the posterior branches.

I take this opportunity to point out that the terms like upper lateral tubercles ("tubercules latéraux supèrieur") in Nikchitch's terminology, or Roman's use of "interne", "médian", and "externe" tubercles (Roman, 1938, p. 425–426) are rather confusing from the standpoint of morphology. As a matter of fact, in *Cheloniceras* and *Acanthohoplites* the position of given tubercles does not interfere with their rank in the series (*i.e.*, umbilical, lateral, or ventral). In the early stages of the *Cheloniceras martini* group, apparently in the earlier stages of the *Cheloniceras cornuelianum* group, and in the innermost coronate whorls of *Acanthohoplites*, the lateral tubercles *first appear at the peripheral margin* and only with the growth of the shell they gradually *pass on the flank*. Because of this, the lateral tubercles may be quite arbitrarily termed as "inferior", "median", or "superior" depending on when and if the umbilical and ventral tubercles make their appearance.

The essential feature associated with the position of true ventral tubercles is seen in that they, if present at all, are invariably *above* the branching point of the primary ribs. The branching, therefor, takes place either at the lateral tubercles (whether they are at the midflank or at the peripheral margin), or at the umbilical tubercles, or at the intermediate points.

It may be argued whether or not a separation of the Cheloniceratidae into "Cheloniceratidae cornuelianinae" and "Cheloniceratidae martiniinae," as suggested by the two groups of Nikchitch, is valid. In the matter of the number of tubercles in the first group it should be noted that d'Orbigny's illustration of *C. cornuelianum* shows, in the earlier part of the last whorl, the angulation and bulging of the ribs as they pass from the flank on the broadly rounded peripheral area and also an appreciable thinning of ribs on the venter between the bulges (d'Orbigny, 1840, Pl. 112, fig. 2). In certain specimens of *C. cornuelianum* from Transcaspian Region, Sinzow (1906, p. 159) noticed the presence of bullate swellings on the ribs between the peripheral margin and the siphonal line below the diameter of 22 mm. and called attention to the similar bullae in the specimen originally placed by d'Orbigny (1840, Pl. 58, fig. 9) with *Ammonites martini*. It may be mentioned in this connection that later Spath was also inclined to place the smaller specimen of d'Orbigny's prototypes with the *C. cornuelianum* rather than with the *C. martini* group of Nikchitch (Spath, 1930, p. 451; 1931, p. 655). Kazansky too observed ventral bullae in young specimens of *C. cornuelianum* from Daghestan (Kazansky, 1915, p. 63).

Noting further Nikchitch's remark (1915) that in the *C. cornuelianum* group (though not in d'Orbigny's type) the primary ribs may branch from the umbilical tubercles in the adult stage, it cannot be accepted with certainty that the branching of the primaries in the *C. martini* group is restricted to the point of lateral tubercles. It is apparent that even in d'Orbigny's type specimen the primary ribs branch from the lateral tubercles only in the young stage, whereas with the growth of the shell the branching point shifts toward the umbilical edge (d'Orbigny, 1840, Pl. 58, fig. 7).

Probably a more valid feature in the development of the Cheloniceratidae is the rate in succession of the coronate-angular-round stages of the shell with the corresponding subtrapezoidal-polygonal-circular whorl section. In the *Cheloniceras cornuelianum* group the angular stage is all but eliminated, whereas in the *Cheloniceras martini* group it attains exaggerated proportions.

The typical *Acanthohoplites* species pass very early through the coronate stage, so that certain characteristics of the mature *Cheloniceras* shell, like the depressed peripheral area, are restricted to the innermost whorls. In others, the round stage, with a circular or oval whorls section, is accelerated, thus decidedly limiting the angular stage to smaller diameters, or even eliminating it altogether.

The *Acanthohoplites* species most resembling *Cheloniceras* are those in which, as in *A. bigoureti* (Seunes, 1887, p. 566, Pl. 14, figs. 3, 4a, 4b), the cheloniceran ornamentation—*i.e.*, the simultaneous presence of strong umbilical and lateral tubercles on the primary ribs, with the branching point of the latter at the lateral tubercles—is retained at larger diameters. In the majority of *Acanthohoplites* species, however, the lateral and umbilical tubercles *do not develop at the same stage of growth, and the thickening of primary ribs at the umbilical edge takes place with the passing of lateral tubercles from the peripheral margin to the midflank.* The umbilical bullae become stronger when the lateral tubercles cease.

Quite apart are species like *Acanthohoplites hesper*, sp. nov. (Pl. 20, figs. 1-6), with a more or less circular and oval whorl section, in which the primary ribs are ornamented at the peripheral margin with lateral tubercles in the early growth, but the latter migrate early to the midflank and at larger diameters the branching point of the former rapidly passes to the umbilical bullae. *Acanthohoplites campichei* (Pictet and Renevier, 1858, Pl. 2, fig. 2) probably also is related to this group.

A reliable corroborant for discrimination between the *Cheloniceras* and *Acanthohoplites* species with similar ornamentation is the suture. In the suture of *Acanthohoplites* the external lobe is long and narrow, with parallel sides. The bipartite external saddle changes considerably during the individual development. In the early whorls it is long and oval, and its outer part is broader than the inner part. At the early maturity, this saddle attains a perfect bifid symmetry in its anterior part but posteriorly its inner side usually slants inward while the outer side remains straight. The first lateral lobe is more or less symmetrically trifid, with the sides either parallel or diverging anteriorly. The first lateral saddle is always broad and symmetrically bifid, with subparallel or parallel sides. The second lateral lobe is small and trifid. The suture is cut by the umbilical seam at the bipartite second lateral saddle. Examples in shells of smaller diameters: Jacob, 1905, p. 408, figure 3; this paper, Figures 8, 10, 11, 13 of Plate 19; and Figure 6 of Plate 20. Examples in shells of larger diameters: Sinzow, 1908, Figures 15, 16 of Plate 4; this paper, Figures 22, 23 of Plate 19; Figure 16 of Plate 20; and Figure 4 of Plate 25.

In the dependent genus *Colombiceras* Spath (d'Orbigny, 1840, Pl. 59, fig. 3) the suture is essentially acanthohoplitan. The suture in the genotype of *Hypacanthohoplites* Spath (Fritel, 1906) has not been figured or discussed. In *Ammonites milletianus* d'Orbigny (1840, Pl. 77, fig. 3) the trifid first lateral lobe is less symmetrical than in *Acanthohoplites* and the saddles are unsymmetrical. This suture, however, belongs to neither of d'Orbigny's two prototypes.

The acanthohoplitan type of suture is strikingly different from the suture of *Cheloniceras* (d'Orbigny, 1840, Pl. 58, fig. 10; Nikchitch, 1915, p. 12, fig. 1; p. 39, fig. 6) in which the external saddle is large and oval but not symmetrically bifid, the first lateral saddle is much shorter than the external saddle, and the first lateral lobe is very wide and divided by a smaller saddle than the external and first lateral saddles. Spath (1931, p. 654) refers to the width of the first lateral lobe of *Cheloniceras* as "enormous." It may be argued whether the first lateral lobe of *Cheloniceras* is correctly interpreted. The fact remains, however, that of the three major saddles of the cheloniceran suture the inner and outer saddles are much larger than the middle one.

Acanthohoplites Sinzow, 1908.

GENOTYPE: *Parahoplites aschiltaensis* ANTHULA (1899, p. 117, Pl. 10, fig. 3a). Holotype by designation.

In establishing his genus *Acanthohoplites* Sinzow (1908, p. 457-458, 478) emphasized the following characteristics as significantly different from those of *Parahoplites* Anthula: (1) presence of tubercles on the ribs, (2) resemblance to "*Douvilleiceras*," that is to *Cheloniceras* in the present interpretation, and (3) essentially symmetrical first lateral lobe in the suture. Sinzow did not specify the genotype of *Acanthohoplites*. Since he regarded *Acanthohoplites aschiltaensis* (Anthula) as the "central form" of his genus, there can be little doubt that this species was meant to be the genotype.

The immediate difficulty arises from the fact that Anthula figured four specimens under *"Parahoplites aschiltaensis,"* all different from one another. The large example in Figure 1 of Plate 11, of Anthula's paper (1899) which shows only one half of the penultimate and last whorls, and has primary ribs with long forward convexities in the upper two thirds of the flank, probably is more related to *"Parahoplites treffryanus"* Anthula (1899, p. 115, Pl. 8, figs. 6a–6c, *non Ammonites treffryanus* Karsten, 1856, p. 109, Pl. 4, figs. 1a–1b) than to the other specimens here discussed.[32] The specimen figured by Anthula in Figure 4 of Plate 10, which shows the cross-section of three last whorls, is ornamented with strong umbilical bullae or elongate tubercles in the ultimate whorl and is, therefore, different from the rest of the illustrated specimens.

Of the remaining two specimens of Anthula's original assemblage, the one he figured in Figures 2a–2b of Plate 10, exhibits the lateral and apertural views; for the other, Figs. 3a–3b of the same plate, the lateral view and the suture are given.

Regarding the first of these two forms Kazansky (1914, p. 68) has already pointed out the discrepancy between Anthula's text and his drawings. According to the description, there are 45 ribs in the final whorl of this specimen, of which 20 are tuberculate primary ribs with from two to three secondary ribs between every two primaries. Consequently, the least possible number of ribs at the given diameter of 70 mm., 19 by 3 plus 1, ought to be 58 and not 45. The important fact is, however, that this specimen, contrary to Anthula's statement in the text, has not a single tuberculate rib on the last whorl.

The example of 45 mm. diameter (Anthula, 1899, Pl. 10, fig. 3a) has from one to three secondary ribs between every two primaries, but the number of the latter on the outer whorl is only 11. This form has strong primary ribs which bifurcate on the last whorl at conical lateral tubercles. If these two specimens of 45 and 70 mm. diameter were identical, at least two primary ribs branching from the lateral tubercles should be visible in the larger form. It is evident that Anthula's description of *"Parahoplites aschiltaensis"* was based on the smaller specimen of 45 mm. diameter which is here designated as the lectotype of *Acanthohoplites aschiltaensis*.

The greatest number of *Acanthohoplites* species, for the first time recognized as congeneric[33] and critically described, is in Sinzow's monograph on the Cretaceous ammonites of Mangyshlak and the Caucasus. Unfortunately, from the nature of collecting, Sinzow often was unable to associate confidently larger forms with the smaller specimens which undoubtedly represented their younger stages; in one case he described two fragments of the same specimen as two different species and did not discover his mistake until after publication of the paper. Discrimination of *Acanthohoplites* species based on presence or absence of the umbilical tubercles, and the nature of the whorl section and of the ribs, is invalid unless a reference to the relative stage of the shell-growth is given.

Some of the species retain into maturity the coronatiform and cheloniceran whorl shape with bituberculate ribs, or ribs branching from the strong lateral tubercles, as in the *A. bigoureti*-*A. abichi* group. In others, the untuberculate ribs and a higher whorl section are acquired early, as in *A. aschiltaensis* (Anthula), or the whorl section becomes subcircular in the early mature stage, as in *A. evolutus* Sinzow and *A. teres*, sp. nov. Further, within the last two groups species with closely or sparsely spaced ribs may be distinguished along with the change in tuberculation attained during the development. In still another group the tuberculation is markedly reduced, as in *A. nolani* (Seunes).

It has been noted elsewhere (Kazansky, 1914, p. 51) that *Acanthohoplites* essentially differs from *Cheloniceras* in the absence of peripheral attenuation of the ribs and a less inflated shell. This is true only for the mature stage of *Acanthohoplites*, which in early youth has all the essential characteristics of the cheloniceran shell. It is significant, however, that the early whorls of *Acanthohoplites* are comparable to the adult and not to the young whorls of *Cheloniceras*. Depressed ribs on the peripheral area are not observed in young stage of the latter genus (d'Orbigny, Pl. 112, fig. 3).

The ornamentation of *Acanthohoplites* differs from that of *Parahoplites* not only in the presence of tuberculate ribs. Two characteristic features of parahoplitan ornamentation: the region of "crowded" ribs on the flank of the young whorls, and the forward sweep of the ribs on the peripheral

---

[32] Sinzow (1908, p. 48) compared this specimen to his *A. laticostatus* and to *A. tobleri* (Jacob).

[33] It is interesting that Burckhardt (1906, p. 191-192) noticed earlier than Sinzow Anthula's inconsistency in placing the untuberculate (*Parahoplites*) and the tuberculate (*Acanthohoplites*) species in one genus. Burckhardt, however, did not exploit the possibilities of his observation.

area, are not observed in the typical species of *Acanthohoplites*. Moreover, as I have mentioned elsewhere, in general the succession of ribs in the *Parahoplitinae* is from flexed to radial, whereas in the Acanthohoplitinae this order of succession is reversed.

One of the outstanding characteristics of the genus is the nature of the first lateral lobe, which makes the entire suture of *Acanthohoplites* so radically different from the sutures of *Parahoplites* and *Cheloniceras*.[34] The nearly symmetrical, sometimes entirely symmetrical, first lateral lobe of this genus has parallel or slightly diverging sides with less prominent lobules, and about three strongly developed branches at the bottom, which makes the lobe appear as a trident with a broad foundation or, conversely, as a mediaeval tower with exaggerated merlons (Pl. 19, figs. 8, 23; Pl. 20, fig. 16; Pl. 25, fig. 4). In mature stage of some species this lobe becomes markedly broad in its anterior part. It should also be mentioned that in *Acanthohoplites* the bifid symmetry of the external saddle is less perfect and is attained in later stages of development than in the Parahoplitinae. Such sutures are observed in many *Acanthohoplites* species described by Sinzow from Mangyshlak and the Caucasus, and in the material collected from the Cretaceous strata of southeastern Arizona.

The described and discussed species that fall within the concept of *Acanthohoplites* as originally interpreted by Sinzow may conveniently be separated into eight principal groups. The first two groups are closely allied. The representatives of the groups 7 and 8 are raised here to the rank of genera.

(1) Group of *Acanthohoplites bigoureti-abichi*, in which the early coronatiform shell, with a more or less broad peripheral area, very gradually develops a round or oval whorl section. The straight uni- or bi-tuberculate ribs, branching either from the lateral tubercles or from the umbilical bullae, are retained in the advanced stages of growth, and the unbranching and untuberculate ribs in the mature whorls remain straight or are very little flexed. Examples: *A. bigoureti* (Seunes, 1887, p. 566, Pl. 14, figs. 3, 4a–4b); *A. abichi* (Anthula, 1899, p. 118, Pl. 9, figs. 2a–2b).

(2) Group of *Acanthohoplites aschiltaensis*, in which the coronatiform shell is restricted to the youngest stages, and the round or oval whorl section is established early. The straight primary ribs, branching from the lateral tubercles at the peripheral margin or the midflank, or bifurcating from the umbilical edge, are superseded in early maturity by flexed unbranching but tuberculate ribs. In the fully mature whorls the ribs are unbranching and untuberculate. Examples: *A. aschiltaensis* (Anthula, 1899, p. 117, Pl. 10, figs. 3a–3b); *A. schucherti*, sp. nov., (Pl. 19, figs. 1–13).

(3) Group of *Acanthohoplites berkeyi*. Little involute species with slowly increasing whorl height and moderately inflated flanks; in adult stage with coarse, sparsely set and slightly flexed primary ribs which appreciably broaden in the peripheral region with tendency toward flattening on the venter. The primary ribs branching from the lateral tubercles at the peripheral margin and bifurcating at the midflank are present in the successive early stages. Example: *A. berkeyi*, sp. nov., (Pl. 19, figs. 14–16). *A. tobleri* (Jacob, 1906, p. 11, Pl. 2, figs. 4a–5b, only) probably also belongs in this group.

(4) Group of *Acanthohoplites impetrabilis*. Little involute species with flattened flanks and sparsely spaced, almost straight, primary ribs which are slightly bullate at the umbilical edge. In the development of earlier whorls the branching point of primary ribs rapidly passes from the lateral tubercles at the peripheral margin to the umbilical bullae. Examples: *A. impetrabilis*, sp. nov., (Pl. 19, figs. 17–20); *A. erraticus*, sp. nov., (Pl. 19, figs. 21–23).

(5) Group of *Acanthohoplites teres*. Very little involute species with a round whorl section, broad venter, deep umbilicus, and sparsely set straight or slightly flexed unbranching primary ribs. Single secondary ribs occur with intervals between the primaries, but often are missing, especially on the last whorl. Strong conical lateral tubercles on the primary ribs at the peripheral margin of young whorls change in the early part of the last whorl to elongated crests placed lower on the flank. Primary ribs do not develop umbilical bullae at any stage. Examples: *A. teres*, sp. nov., (Pl. 20, fig. 7). The densicostate and less evolute *A. evolutus* Sinzow (1908, p. 492, Pl. 4, figs. 21–22), with single secondaries between branching primary ribs, is the nearest comparable form that I know.

---

[34] The suture of *Cheloniceras? gottschei* from Zululand, restored by Spath (1921, p. 312, Pl. 26, fig. 1d), is rather acanthohoplitan than cheloniceran. The difference in length between the narrow external saddle and the wider bipartite first lateral saddle, with parallel sides, is not as impressive as in *Cheloniceras*. Note also the tendency toward a strong tripartition in the first lateral lobe.

(6) Group of *Acanthohoplites hesper*. Species with round to oval whorl section and numerous secondary ribs between the branching primaries. The branching of the primaries gradually passes from the lateral tubercles at the peripheral margin and the midflank in the young whorls to the strong oblong umbilical bullae in the mature shell. Example: *Ac. hesper*, sp. nov., (Pl. 20, figs. 1-6). *Ammonites campichei* Pictet and Renevier (1858, p. 25, Pl. 2, figs. 2a-2c, non *A. campichei* Pictet and Renevier in Pictet and Campiche, 1860, p. 258, Pl. 37, figs. 1a-1b, non *Parahoplites campichei* Sinzow, 1908, p. 460, Pl. 1, figs. 4-7) probably belongs in this group.

(7) Group of *Acanthohoplites immunitus*. In the species of this group the tuberculation is greatly reduced and restricted to the younger stages of growth, at which the closely set primary ribs are slightly bullate at the umbilical edge and also bear incipient or very small lateral tubercles which migrate from the peripheral margin to the midflank with the growth of the shell. The adult whorls, with a somewhat flattened peripheral area, are densicostate, the primary ribs develop well pronounced shallow sinuosities in the umbilical and peripheral parts of the flank, and a forward convexity about the midflank. A new genus *Immunitoceras* is proposed for the species of this group. Examples: *I. immunitum*, sp. nov., (Pl. 20, figs. 8-15); *I. nolani* (Seunes, 1887, p. 564, Pl. 13, figs. 4a, 4b); *I. nolani* with varieties *pigmaea* and *crassa* (Sinzow, 1908, p. 503, Pl. 8, figs. 2-5 and 11-13). *Immunitoceras? uhligi* (Anthula, 1899, p. 114, Pl. 10, figs. 1a-1b) probably also belongs in this genus.

(8) Group of *Acanthohoplites meridionalis*. In the development of the shell in the species assigned to this group the early coronatiform stage, with depressed peripheral ribs, and the stage with a reniform whorl section, are either greatly reduced or altogether eliminated. With the maturity the successively round and oval whorl sections change to polygonal, at first through the rapid increase of the whorl height and the development of unbranching primary ribs ornamented with strong conical lateral tubercles at the midflank, and later also through the thickening and angulation of the primary ribs at the umbilical edge and the peripheral margin, without, however, the formation of individualized bullae. For such species, with unituberculate and unbranching primary ribs, and a polygonal whorl section, the generic name *Paracanthohoplites* is proposed. Examples: *P. meridionalis*, sp. nov. (Pl. 21, figs. 1-7); *P. multispinatus* (Anthula, 1899, p. 119, Pl. 10, figs. 5a-5c).

With all modifications in the development of the shell and ornamentation the characteristic acanthohoplitan suture varies little in the species under discussion, as is seen in Sinzow's photographic reproductions of the material from Mangyshlak and the Caucasus and in my collection from southeastern Arizona. It may seem rather peculiar that the sutures of *Acanthohoplites aschiltaensis* and *Paracanthohoplites multispinatus*, as drawn by Anthula (1899, Pl. 10, figs. 3b and 5c; *see* also Roman, 1938, p. 350), deviate from the typical acanthohoplitan suture, as for instance in Plate 25, figure 4. The first lateral lobe of the first species is represented by Anthula as asymmetrically bifid, a feature that does not occur in any other species of *Acanthohiplites*, and the same lobe of the second species is drawn very narrow in its terminal part. In my opinion these sutures were not drawn accurately enough by Anthula, or else they were taken from unsatisfactorily preserved specimens.

Acanthohoplites schucherti Stoyanow, sp. nov.

(Plate 19, figures 1-13)

MEASUREMENTS:

	Holotype No. 91199	Paratype No. 91675
Diameter	29	33
Greater radius	17	19
Lesser radius	12	14
Height of last whorl	11	14
Thickness of last whorl	10	14
Width of umbilicus	10	14

The early shell of this species is coronatiform or cheloniceran in appearance. At the diameter of 3 mm. the whorl section is subtrapezoidal with the greatest thickness of the whorl a little below the rounded peripheral margin, and with the flanks moderately diverging toward the periphery. Lateral tubercles, already present at this diameter, are situated slightly below the peripheral margin. They appear as low but prominent nodes with a round outline and occupy almost the entire upper third of the flank. Primary ribs are not distinctly indicated on the flanks except for very blunt, uneven, barely perceptible prominences the radial nature of which is rather guessed than observed and which rapidly merge into the lateral tubercles above. On the other hand, the ribbing is clearly developed on the peripheral area: the stronger primary ribs connect the lateral tubercles across the venter and alternate with the weaker secondary ribs which extend between the untuberculated parts of the peripheral margins. All these ribs are straight.

Between diameters of 5 and 10 mm. the straight primary ribs are fully developed on the flank; slender at the umbilical seam, they very rapidly increase in thickness and terminate in robust conical lateral tubercles placed exactly at the peripheral margin. From this point the primary ribs bifurcate on the venter into the weaker and slightly bent forward anterior branches and the stronger and straight posterior branches. At this stage the secondary ribs encroach from the venter upon the upper third of the flank, and are of the same strength and attitude on the peripheral area as the anterior branches of the primary ribs. On the venter all the ribs are slightly depressed, and the posterior branches of the primary ribs appreciably thicken on the sides of the siphonal line. These swellings are regarded here as incipient ventral bullae. Indeed, although these ribs are low on the marginal part of the peripheral area above the point of branching from the lateral tubercles, they noticeably increase in thickness on the venter proper and form blunt elongate bullae which cease completely before the siphonal line is reached, not unlike the similar bulges on the periphery of *Cheloniceras cornuelianum* (d'Orbigny, 1840, Pl. 112, fig. 2). The whorl section remains subtrapezoidal between the indicated diameters. The umbilical wall is somewhat convex in its upper part and slightly concave in the lower part. In the suture, the external lobe has parallel sides. The external saddle, suboval in outline, is placed entirely on the peripheral area, it is unequally bifid with the outer part larger than the inner part. The first lateral lobe, wide anteriorly and indefinably trifid, is at the peripheral margin and coincides with the lateral tubercle. The broad and bifid first lateral saddle is below the lateral tubercle and occupies the greater part of the flank, it is separated from the umbilical seam by the small second lateral lobe and the small second lateral saddle.

At a diameter of 12 mm. the whorl section is subreniform, and shortly after becomes, except for the antisiphonal side, transversely ovate with gently curved peripheral area and flanks. At this stage the ribs are less closely spaced and the conical lateral tubercles are very prominent. The umbilical wall is steep.

About the diameter of 18 mm. the whorl section attains almost round outline. The primary ribs, thin and high in the umbilical wall, bend nearly at right angle over the umbilical edge as they pass on the flank.

Above the diameter of 18 mm. the lateral tubercles pass below the peripheral margin, and accordingly, the primary ribs branch slightly above the midflank. There is only one secondary rib between the bifurcating primaries. All the ribs are of the same thickness on the peripheral area and pass over the siphonal line without attenuation.

Beginning with diameters of 22–25 mm. the relation between the branching point of the primary ribs and the lateral tubercles is less regular, and the bifurcation takes place at, near, or appreciably below the tubercles. The ribs are more flexed on the flank and distinction between the primary and secondary ribs is less definite. The whorl section changes from round to subcircular, and the peripheral area is somewhat flatter and narrower. Modifications in the ribbing at 25 mm. and larger diameters are of the following relations: (1) The primary rib loses the lateral tubercle and its point of bifurcation passes down the flank closer to the umbilical edge. (2) With a similar migration of the branching point the elongate and crest-like tubercle is retained in one of the two branches higher on the flank. (3) One branch is separated from the primary rib to form a new secondary rib and the lateral tubercle remains on the primary rib. (4) The tubercle is retained by both the primary rib and by the separated branch.

The suture at this and larger diameters is typically acanthohoplitan. The external lobe is long, narrow, and with parallel sides. The external saddle is symmetrically bifid anteriorly and with subparallel sides. The symmetrically trifid first lateral lobe is wide anteriorly. The first lateral saddle is symmetrically bifid and with parallel sides. The small second lateral lobe is trifid. The bipartite second lateral saddle is cut by the umbilical seam.

The ornamentation in adult whorls of *Acanthohoplites schucherti* is essentially of the same nature as in *A. aschiltaensis* (Anthula, 1899, Pl. 10, fig. 3a). In the latter species the lateral tubercles at the branching point of the primary ribs are retained at larger diameters of the shell, and the number of secondary ribs between the primaries is greater than in the Arizona species. A similar migration of the branching point and of the lateral tubercles is also observed in *A. bigoureti* (Seunes, 1887, Pl. 14, figs. 3, 4a; Sinzow, 1908, Pl. 6, fig. 5) and to a lesser extent in *A. abichi* (Anthula, 1899, Pl. 9, fig. 2a; Sinzow, 1908, Pl. 6, figs. 1, 2, 3).[35]

TYPE: Holotype: No. 91199. Paratype: No. 91675. Metatypes: Nos. 91193, 91194, 91195, 91196, 91197, 91198 (not illustrated).

OCCURRENCE: All types collected from Quajote member, division 3d, of Lowell formation, Ninety One Hills.

Acanthohoplites berkeyi Stoyanow, sp. nov.

(Plate 19, figures 14-16)

MEASUREMENTS:

	Holotype No. 92230
Diameter	47
Greater radius	27
Lesser radius	20
Height of last whorl	18
Thickness of last whorl	17
Width of umbilicus	15

The shell of this species is septate to a diameter of 40 mm. and little involute, with a subcircular whorl section; its flanks are moderately inflated, and the ribs are sparsely set. The early whorls, at diameters of about 5-6 mm., are more densicostate, their primary ribs are straight on the flank and bifurcate on the venter from the lateral tubercles placed at the peripheral margin. This nature of ornamentation changes very early, and at a diameter of 15 mm., as observed in a number of specimens, the ribs already are sparsely spaced. At this stage the branching of the primary ribs at the peripheral margin stops, and these ribs either become simple and retain a strong lateral tubercle high on the flank, or lose the tubercle but retain the branching, the point of which passes down the flank. Tuberculation on single primary ribs ceases early however, whereas the untuberculate bifurcating ribs are sporadically present to 40 mm. diameter. It should be noted that there are varietal forms in which the young shell, with somewhat flattened flanks and venter, is densicostate to a diameter of 15 mm. (Pl. 19, fig. 16). The secondary ribs, usually one between each pair of primaries, gradually encroach with development from the venter onto the flank. Some of these ribs reach the midflank in the penultimate whorl and the early part of the last whorl but are shorter again on the living chamber. In the last whorl the primary ribs are slightly flexed and somewhat attenuated at the midflank, but increase in thickness toward the peripheral margin. On the venter the ribs are strong, straight, and rounded. Shortly before the living chamber is reached, the ribs develop a backward slope on the peripheral area which in the terminal part of the shell attains exaggerated proportions. The suture of this species is typically acanthohoplitan. The first lateral lobe is broad with parallel sides, in which the lobules are small, and with three long, strong, trident-like, branches at the bottom.

Acanthohoplites berkeyi resembles *A. tobleri* (Jacob, 1906, p. 11, Pl. 2, figs. 4a-5b, only). The latter species retains denser ribbing at larger diameters, has flatter ribs on the venter, and in younger stage has not only bifurcating but also trifurcating primary ribs. Jacob did not mention either the presence or the absence of tubercles in the inner whorls of his species; nor is there any reference to tuberculation in *A. tobleri* described by Sinzow (1908, p. 486, Pl. 5, figs. 14-15) from the Black Sea coast of the Caucasus, which form differs both from Jacob's and my species in straighter and denser ribbing of the last whorl. On the other hand, tuberculate primary ribs are present in *A. tobleri* var. *discoidalis* Sinzow (1908, p. 486, Pl. 5, figs. 17-20), a form from the northern Caucasus, that has a

---

[35] Sinzow (1908, p. 488) was inclined to regard *"Parahoplites" bigoureti* of Anthula (1899, p. 117, Pl. 13, figs. 2a-2c) as a variety of *A. abichi*.

high and compressed whorl section throughout growth. Of the two related forms from Colombia described by Riedel (1937, p. 49–51, Pl. 8, figs. 20–22 and 23–24) as *Colombiceras tobleri* var. *discoidalis* Sinzow and *C.* aff. *tobleri* Jacob the latter form has primary ribs branching from lateral tubercles, whereas the ribs of the former lack both the branching and the tuberculation.

Jacob compared his species to *A. aschiltaensis* (Anthula), *Ammonites crassicostatus* d'Orbigny (1840, p. 197, Pl. 59, figs. 1–4′), and *A. treffryanus* Karsten (1856, p. 109, Pl. 4, figs. 1a–1b). The two latter species usually are placed under *Colombiceras* in the current literature. The genus *Colombiceras* was established by Spath (1923a, p. 64; *compare* 1921, p. 317–318) for the forms with peculiarly flattened ribs both on the flanks and on the venter, and with a well-differentiated truncation of the whorl section, as in *Colombiceras crassicostatum* (d'Orbigny, 1840, Pl. 59, figs. 1 and 2, only). The principal difficulty in connection with this designation is in the fact that it is not known whether the designated holotype possesses tuberculate ribs at any stage of the growth. The smaller prototype of *Ammonites crassicostatus*, which according to d'Orbigny (1840, p. 197, Pl. 59, figs. 4–4′) is a young form of his species and has primary ribs that bifurcate from the lateral tubercles at the midflank, was not included in Spath's genotype. Spath (1923a, p. 64) placed *Colombiceras* in the family Cheloniceratidae. In a later publication he (Spath, 1930, p. 441) considered *A. tobleri* as a transitional form between *Parahoplites* and *Colombiceras*, which barely is possible if the latter genus has acanthohoplitan ornamentation in the early youth and if the suture drawn by d'Orbigny (1840, Pl. 59, fig. 3) for his *Ammonites crassicostatus* belongs to the genotype. Status of *Colombiceras* is discussed elsewhere in this paper.

TYPE: Holotype: No. 92230. Paratypes: Nos. 92231, 92232, 92233, and the varietal form No. 92249. (No. 92249 is illustrated in Pl. 19, fig. 16).

OCCURRENCE: All types collected from Quajote member, division 3b, of Lowell formation, Ninety One Hills.

Acanthohoplites impetrabilis Stoyanow, sp. nov.

(Plate 19, figures 17-20)

MEASUREMENTS:

	Syntype No. 92228	Syntype No. 92198
Diameter	40	25
Greater radius	24	15
Lesser radius	16	10
Height of last whorl	16	11
Thickness of last whorl	16	11
Width of umbilicus	15	13

Of two identical syntypes, No. 92228 and No. 92198, the latter has been dissected to illustrate the penultimate whorl.

The essential characteristics of this species: flat flanks, very little arched peripheral area, involution not surpassing one fourth of the whorl, subradial ribs, and subquadrate whorl section, are established very early, though the flatness of the flanks increases with growth, while the ribs become slightly flexed.

In the earlier whorls, the primary ribs gradually thicken on the flank from the umbilical edge, are radial, and bifurcate at the peripheral margin from conical lateral tubercles into anterior and posterior branches which pass over the venter without attenuation. The secondary ribs do not occur on the flank in early stages, as a rule, although in some specimens one or two secondary ribs appear rather early. The lateral tubercles cease at a diameter of 12–13 mm. Simultaneously the primary ribs begin to bifurcate at the umbilical edge. At first they are weak at the point of bifurcation, but at a diameter of 16 mm. umbilical bullae are formed at the branching point. The secondary ribs appear regularly at about the midflank between the primaries. With further growth, the umbilical bullae are supplanted by thinner crests, the anterior branches of the primary ribs are less traceable to the point of branching, and occasionally there are two secondary ribs between the primaries.

Above a diameter of 20 mm. the anterior branches of the primary ribs become weaker, sometimes are detached, and in this way additional secondary ribs appear lower on the flank than at lesser diameters. The single primary ribs are thicker at the umbilical edge and the peripheral margin. Some of the primary ribs "split" with disruption of continuity, like in *Acanthohoplites erraticus*, sp. nov. All the ribs are slightly flexed on the flank and equally strong on the venter, over which they pass without attenuation.

The suture, with a stout and symmetrically trifid first lateral lobe, is typically acanthohoplitan from the earliest stages.

The species of *Acanthohoplites* with radial ribs and more or less flat flanks, which were described by Sinzow (1908, p. 490, Pl. 4, fig. 7 and p. 499, Pl. 4, figs. 15-17) from Mangyshlak and the Caucasus as *A. subangulatus* Sinzow and *A. trautscholdi* (Simonovitsch), are more densicostate and have a higher whorl section.

TYPE: Syntypes: Nos. 92228 and 92198.

OCCURRENCE: Both syntypes were collected from Quajote member, division 3b, of Lowell formation, Ninety One Hills.

Acanthohoplites erraticus Stoyanow, sp. nov.

(Plate 19, figures 21-23)

MEASUREMENTS:

	Holotype No. 92500
Diameter	45
Greater radius	27
Lesser radius	18
Height of last whorl	17
Thickness of last whorl	16
Width of umbilicus	18

The shell of this species is of medium size, very little involute, with the whorls slowly increasing in height and moderately inflated, septate to 32 mm. diameter. The venter is very gently arched, the flanks are slightly covergent in the last whorl, the greatest thickness of the living chamber is at the umbilical edge.

In the inner whorls the primary ribs are radial, sparsely set, and bear strong conical lateral tubercles placed at the peripheral margin. The secondary ribs are absent on the flank or are indicated in places by barely perceptible elevations.

At a diameter of 20 mm. the primary ribs become stronger as they pass from the umbilical wall to the flank over the umbilical edge. The lateral tubercles are slightly above the midflank. The secondary ribs, one between each pair of the primaries, are distinct and situated in the upper third of the flank.

In the last part of the penultimate whorl the lateral tubercles change from conical to crest-like, they cease in the early part of the last whorl. At this stage the primary ribs, straight or slightly flexed, are thicker at the umbilical edge and the peripheral margin, and on the venter.

Before the body chamber is reached the distribution and attitude of the secondary ribs becomes markedly irregular. As the lateral tubercles cease, the secondary ribs appear in twos for the first time. Next, there is again one secondary rib between the primaries. Between the two following primary ribs the posterior secondary rib is normal, whereas the anterior secondary rib is convergent to the primary rib in front of it and joins it at the peripheral margin. This compound primary rib passes over the venter, but on the opposite flank it continues as a secondary rib, while the corresponding primary rib, originating at the umbilical edge, does not reach the peripheral margin. This arrangement is repeated three times on the body chamber. Between such sets are ordinary primary ribs with one secondary rib between every pair. One of the primary ribs bifurcates at the umbilical edge, but the posterior branch becomes an independent secondary rib on the opposite flank. As a result, the ribbing in the terminal part of the last whorl is very irregular and on the flank the primary ribs have an interrupted appearance.

The suture of this species is characteristically acanthohoplitan with bifid saddles and a strongly trifid, anteriorly wider, first lateral lobe.

Acanthohoplites erraticus somewhat resembles *A. impetrabilis* but disagrees with it in more inflated flanks, more round whorl section, and conspicuously stronger and irregularly distributed secondary ribs.

"Interrupted" primary ribs similar to those of *Ac. erraticus* have been observed in *Arcthoplites jachromensis* (Nikitin, 1888, p. 57, Pl. 4, figs. 1-2), in *Cheloniceras buxtorfi* (Jacob, 1906, p. 15, Pl. 1, fig. 10a), and in certain forms related to the *Ammonites milletianus* group.

TYPE: Holotype: No. 92500. Paratype: No. 92501 (fragmental and not illustrated).

OCCURRENCE: Holotype and paratype were collected from Quajote member, division 3b, of Lowell formation, Ninety One Hills.

Acanthohoplites teres Stoyanow, sp. nov.

(Plate 20, figure 7)

MEASUREMENTS:

	Holotype No. 92229	Paratype No. 92200
Diameter	34	27
Greater radius	19	16
Lesser radius	15	11
Height of last whorl	13	11
Thickness of last whorl	13	11
Width of umbilicus	12	8

The shell of this species is very little involute at lesser diameters and nearly evolute at a diameter of 34 mm. It has a round whorl section and a broad gently arched peripheral area. The umbilicus is deep and the umbilical wall is steep. The whorls increase in height and thickness very gradually. The primary ribs, radial in the young and somewhat flexed in the adult whorls, do not branch on the flank at any stage of development and on the venter of the last whorl.

In earlier stages the straight primary ribs are thin at the umbilical edge, and in the lower part of the flank, but rapidly increase in thickness toward the peripheral margin, at which they bear very strong conical lateral tubercles. It is not known whether in the early whorls the primary ribs branch from the lateral tubercles at the peripheral margin, as is the case in other species of *Acanthohoplites*. With advanced growth the primary ribs are low in the umbilical wall, stronger at the umbilical edge, but do not form the bullae as they pass onto the flank. They are thin and slender at the midflank, and, instead of conical tubercles, bear at the peripheral margin very narrow thorn-like crests, so thin and straight that they do not show well in illustration; with further growth they become weaker and die out at the whorl height of 10-12 mm.

Scarcity of secondary ribs in this species should be noted. Occasionally present in the early whorls, they are so high on the flank as to be almost unexposed. There are only three secondary ribs in the penultimate whorl. Four secondary ribs alternate normally, high on the flank, in the early part of the last whorl. In the later part of the last whorl two short secondaries are present between diameters of 24 and 28 mm., and a longer secondary rib is at the diameter of 32 mm. in the terminal part of the shell. All the ribs cross the peripheral area without interruption, but also without any appreciable increase in thickness.

The suture is not quite satisfactorily preserved. Nevertheless, the typically trifid first lateral lobe and the bifid first lateral saddle are clearly seen in the penultimate and last whorls of the holotype.

This species differs from the more involute *A. evolutus* Sinzow (1908, p. 492, Pl. 4, figs. 21-22) in a considerably lesser number of ribs, scarcity and irregular distribution of the secondary ribs, and absence of branching ribs on the flank.

TYPE: Holotype: No. 92229. Paratype: No. 92200 (not illustrated).

OCCURRENCE: Holotype and paratype collected from Quajote member, division 3b, of Lowell formation, Ninety One Hills.

Acanthohoplites hesper Stoyanow, sp. nov.

(Plate 20, figures 1-6)

MEASUREMENTS:

	Syntype No. 92226	Syntype No. 92227
Diameter	29	23
Greater radius	19	14
Lesser radius	10	9
Height of last whorl	16	10
Thickness of last whorl	14.5	10
Width of umbilicus	8	7

The shell of this species is one third involute, densicostate, with rounded venter and peripheral margins, rather narrow umbilicus, and steep umbilical wall. The greatest whorl thickness is at the umbilical edge. The whorl section rapidly passes with the growth from round to suboval with the flattening of the flanks which slightly converge toward the venter. The ornamentation appreciably changes with the development because the branching point of the primary ribs and the lateral tubercles migrate, the latter are substituted with umbilical bullae, and the secondary ribs gradually encroach upon the flank.

In the earliest observed whorls the primary ribs are sparsely spaced and straight on the flank. At the peripheral margin they bifurcate from strong conical lateral tubercles. At the same time five secondary ribs appear on the venter between the branching primaries. At a whorl-height of 5-6 mm. the lateral tubercles and the branching point of the primary ribs pass down the flank from the peripheral margin, and the secondary ribs extend on the flank from the venter. With the process of growth the point of bifurcation of the primary ribs migrates to the midflank, simultaneously the lateral tubercles become weaker, whereas the primary ribs thicken at the umbilical edge. At a diameter of 9-10 mm., the point of bifurcation is *below* the lateral tubercle which remains at the midflank in the posterior branch of a primary rib not yet strongly bullate at the umbilical edge. Above this diameter, however, the lateral tubercles cease, and at a diameter of 12-13 mm. the primary ribs already bifurcate from strong oblong umbilical bullae near or at the umbilical edge. The number of secondary ribs between each pair of the primaries remains 5 to 13 mm. diameter, and changes consecutively to 4 and 3 at larger diameters. Since the bifurcation of the primary ribs continues to the end of the last observed whorl, care should be taken not to confuse their weaker anterior branches with the secondary ribs. The latter ribs are weak in the lower part of the flank but increase in thickness toward the peripheral margin. In the last whorl all the ribs are slightly flexed on the flank, with the flexuosity accentuated near the peripheral margin, and equally strong on the venter, which they cross without attenuation.

The suture is typically acanthohoplitan. In the earlier whorls the first lateral lobe has diverging sides which gradually change to subparallel and parallel, and develops three strong branches at the bottom in the advanced stages.

Acanthohoplites hesper shows certain resemblance to *Ammonites campichei* Pictet and Renevier (1858, p. 25, Pl. 2, figs. 2a-2c, non *A. campichei* Pictet and Campiche, 1860, p. 258, Pl. 37, figs. 1a-1b, non *Parahoplites campichei* Sinzow, 1908, p. 460, Pl. 1, figs. 4-7. Also not to be confused with *Hyphoplites campichei* Spath, 1925, p. 83). Pictet's statement that all the ribs of his species are continuous with the umbilical bullae is incorrect. The illustration clearly shows that only the primary ribs bifurcate from the umbilical bullae, whereas the secondaries appear higher on the flank and are very slender below. Together with the distribution of secondary ribs and the oblong nature of the umbilical bullae, the ornamentation of this species does not differ materially from that of *Acanthohoplites hesper* if allowance is made for the general attenuation of ribs with age, according to Pictet's observation. The suture of Pictet's species, with a bifid external saddle, a symmetrically trifid first lateral lobe, and bifid first and second lateral saddles, both with parallel sides, is acanthohoplitan. Comparison is handicapped, because the earlier whorls of *Acanthohoplites campichei* are not known and, therefore, no inference can be made regarding the development of ornamentation, and because the shell of the Arizona species is known only to a diameter of 29 mm.

The incomplete specimen described and illustrated by Pictet and Campiche in 1860 under the specific name *campichei* is not conspecific with the holotype from the Rhône, its primary ribs do not form umbilical bullae and the general distribution and branching of ribs is quite different. Later Sinzow (1908, p. 463, Pl. 1, figs. 8-9) identified this fragment with his species *Parahoplites subcampichei*.

The ammonites from Transcaspian Region described by Sinzow as *Parahoplites campichei* (Pictet and Renevier) do not belong to that species and most probably are not congeneric among themselves. While the fragment figured in Figure 7 of Plate 1 of Sinzow's monograph has a nearly symmetrically trifid first lateral lobe and probably is an acanthohoplite, the suture in the large form of his Figure 4 of Plate 1, with an asymmetrical first lateral lobe, is typically parahoplitan. Sinzow emphasized the difference between the strong umbilical bullae of *A. campichei* (Pictet and Renevier) and the long narrow umbilical thickenings in the primary ribs of his assemblage. It may be mentioned in passing, that such primary ribs are indeed quite common in several groups of the Parahoplitinae, including *Sinzowiella*, although they are not conspicuous in *Parahoplites melchioris* (Anthula) and allied genera, like *Kazanskiella* described in this paper. Kazansky (1914, p. 91), noting the difference between the species of Pictet and Renevier and *Parahoplites campichei* as described by Sinzow, suggested that the latter might be a different species with a parahoplitan rather than acanthohoplitan suture. He did not discriminate, however, among the heterogeneous forms of Sinzow's assemblage. Spath (1930, p. 438, fig. c; p. 439) reported the presence of specimens in the Upper Aptian of England that are comparable to two forms illustrated by Sinzow and figured the suture of one of such examples in the British Museum with a large bifid external saddle, an asymmetrically trifid first lateral lobe, and an asymmetrical first lateral saddle, much smaller than the external saddle.

Acanthohoplites pulcher Riedel (1937, p. 43, Pl. 8, figs. 11-14) of Colombia has three to five secondary ribs between the primaries, and some of the latter bear elongate lateral tubercles high on the flank. The umbilical bullae, however, are not developed in this species.

TYPE: Syntypes: Nos. 92226 and 92227.

OCCURRENCE: Both syntypes collected from Quajote member, division 3b, of Lowell formation, Ninety One Hills.

Immunitoceras Stoyanow, gen. nov.

GENOTYPE: *Immunitoceras immunitum* STOYANOW, sp. nov.

This genus is proposed for the densicostate species of the subfamily Acanthohoplitinae in which the ribs are essentially subsigmoidal and the tuberculation is greatly reduced. In *Immunitoceras nolani* (Seunes, 1887, p. 564, Pl. 13, figs. 4a-4b) the closely set ribs branch low on the flank and are not readily differentiable into primaries and secondaries in the early whorls, tuberculation of the ribs is limited to lesser diameters of the shell, and only incipient lateral tubercles are present.

The densicostate forms from Mangyshlak described as *Acanthohoplites nolani* by Sinzow (1908, p. 503, Pl. 8, figs. 1-13) are not conspecific with Seunes' holotype and with each other, which probably accounts for the varietal names "*subrectangulata*," "*pygmaea*," and "*crassa*." In the young whorls of these ammonites the umbilical bullae and lateral tubercles are appreciably developed although Sinzow refers to them as "microscopic" and "atrophied." Renngarten (1926, p. 25) states that the forms from the Cretaceous of the Caucasus are closer to Seunes' holotype than to the specimens from Mangyshlak.

A similarity between *Immunitoceras? uhligi* (Anthula, 1899, p. 114, Pl. 10, figs. 1a-1b) and *I. nolani* (Seunes) has already been noted by Sinzow (1908, p. 498). In this species the ribs branch low on the flank already in the early whorls, in the final whorl the primary ribs are subsigmoidal and the secondaries appear low on the flank. Anthula mentions umbilical bullae on the ribs of the last whorl and also very strong thorn-like tubercles at the umbilical edge of the inner whorls. The latter tubercles, however, are definitely absent in his illustration. If the lateral tubercles are present at all at the peripheral margin of this species, they must be restricted to its earliest whorls, because the bifurcating point of the ribs migrates to the flank very early, and the very first branching rib is untuberculate on the flank. The suture of this species is unknown.

A general tendency toward reduced tuberculation is observed in all species of this genus. Since, however, this reduction varies indefinably, a separation of species based on the degree of reduction alone would not be feasible. On the other hand, it is evident that within the compass of the genus are: (1) species in which the primary and secondary ribs are differentiated in the earlier whorls, and the tuberculation is relatively stronger, as in *I. nolani* (Sinzow, 1908, Pl. 8, figs. 2, 3, 5, non Seunes 1887) and the varieties *subrectangulata* and *crassa* (Sinzow, 1908, Pl. 8, figs. 6, 8, and 11, 13), and in *I. immunitum*, sp. nov., described in this paper (Pl, 20, figs. 8, 13); and (2) species in which the ribs are not distinctly differentiated on the flank of the inner whorls and the tuberculation on the ribs is reduced to its minimum, as in *I. nolani* (Suenes, 1887), *I.? uhligi* (Anthula, 1899), and probably in *I. nolani* var. *pygmaea* (Sinzow, 1908, Pl. 8, fig. 4).

The suture of *Immunitoceras* is acanthohoplitan (*see* Sinzow, 1908, Pl. 8, fig. 1).

Immunitoceras immunitum Stoyanow, sp. nov.

(Plate 20, figures 8–15)

MEASUREMENTS:

	Holotype No. 92563
Diameter	64
Greater radius	42
Lesser radius	22
Height of last whorl	31
Thickness of last whorl	22
Width of umbilicus	21

Involution of the shell in the described species consecutively changes from one half, through one third, to one fourth of the whorl. In the last whorl the flanks are somewhat flattened, the peripheral area is broadly arched, and the umbilical wall is steep. The untuberculate primary ribs form shallow sinuosities in the umbilical and peripheral parts of the flank and a low median convexity. The secondary ribs appear below or at the midflank.

In the earlier stages the whorl section changes from round to oval, and the ribs are not distinctly differentiated on the flank. The primary ribs, however, are indicated already below the whorl height of 5 mm. by the point of bifurcation with a small lateral tubercle at the peripheral margin, while the secondary ribs gradually encroach from the venter upon the flank. At the whorl height of 5 mm. the point of bifurcation abruptly passes to the lower third of the flank but weak lateral tubercles are discernible on the posterior branches of the primary ribs below the peripheral margin, or in the single primary ribs after the separation of their anterior branches which form additional secondary ribs. With the detachment of anterior branches the primary ribs thicken at the umbilical edge and on the venter, and to a diameter of 25 mm. all the ribs are somewhat angular as they pass from the peripheral margin to the peripheral area. This causes a change from the oval to subangular whorl section (Pl. 20, figs. 9 and 10), very similar to that observed in *I. nolani* var. *subrectangulata* (Sinzow, 1908, Pl. 8, fig. 10). However, with further growth and cessation of all tuberculation the whorl section again becomes suboval and oval.

On the last whorl the primary and secondary ribs alternate, but occasionally a secondary rib may be traced to the neighboring primary rib as its derivative. On the venter all the ribs are rounded (not flat as it appears in Pl. 20, fig. 15 because of a faulty illumination).

The suture, preserved only in the inner whorls of the holotype, is characteristically acanthohoplitan.

Immunitoceras immunitum is related to those species of the genus in which the primary ribs with weak tubercles are differentiated in the early stages of growth.

TYPE: Holotype: No. 92563. Paratype: No. 92231.

OCCURRENCE: Holotype and paratype collected from Quajote member, division 3b, of Lowell formation, Ninety One Hills.

Paracanthohoplites Stoyanow, gen. nov.

GENOTYPE: *Paracanthohoplites meridionalis* STOYANOW, sp. nov.

The study of abundant congeneric material from the Lowell formation reveals that certain species of ammonites which were placed by Sinzow (1908) in *Acanthohoplites* do not develop in the early stage growth a coronatiform or cheloniceran shell with a broad depressed venter, which is so characteristic of *Acanthohoplites* in that stage. Instead, their earliest and young whorls are round, with a circular and transversally oval whorl section which changes to polygonal in mature stage. The succession of the whorl section in *Paracanthohoplites* is circular—transversally oval—oval—polygonal—suboval (with a subtabulate venter), unlike the subtrapezoidal—reniform—circular—oval succession of the whorl section in *Acanthohoplites*. Ornamentation of *Paracanthohoplites* from the early to mature whorls consists essentially of unbranching unituberculate primary ribs which alternate with three, two, and one secondary rib. This ornamentation disagrees with that of *Acanthohoplites* in which the branching primary ribs with migrating point of bifurcation are especially diagnostic. The suture of this genus has the same general character as in *Acanthohoplites*. Differences in the development of shell and ornamentation warrant the separation of such species from *Acanthohoplites*. Certain species, also with unbranching and unituberculate ribs, placed in the last genus, like *A. teres* (Pl. 20, fig. 7), have a totally different development of involution, whorl section, and costation. As far as I know, "*Parahoplites*" *multispinatus* Anthula (1899, p. 119, Pl. 10, figs. 5a–5c) is the first described species of the proposed genus. Transitional and deviating forms are discussed in the description of the genotype.

Paracanthohoplites meridionalis Stoyanow, sp. nov.

(Plate 21, figures 1–7)

MEASUREMENTS:

	Holotype No. 91150	Paratype No. 91152
Diameter	33	62
Greater radius	21	36
Lesser radius	12	26
Height of last whorl	17	21
Thickness of last whorl	13	17
Width of umbilicus	10	18

Numerous specimens of this species afford an abundant material for detailed studies. In the first three volutions the shell is smooth with a round whorl section. At the whorl height of 3 mm. the whorl section changes to transversally oval. Simultaneously the first primary rib appears on the venter and terminates in a relatively large and high thorn-shaped lateral tubercle which occupies nearly all space between the midflank and the peripheral margin. Between this stage and the diameter of 8 mm., the flank has seven unbranching primary ribs, each ornamented with a single conical lateral tubercle which is placed close to the peripheral margin, and three secondary ribs between the first and the second, and between the second and the third primary ribs, but only two secondaries between every two following primaries. At first the secondary ribs are weaker on the venter than on the flank, but gain in strength with development. At this stage the ribbing is dense and all the ribs are straight on the flank and on the peripheral area. Because of the sudden increase in the whorl thickness the umbilicus is deep.

Above the diameter of 8 mm. the primary ribs markedly increase in thickness both on the flank and on the venter, the strong conical lateral tubercles migrate to the midflank, and the entire costation acquires a coarse aspect. Three flexed secondary ribs are consistently present between the primaries. At about 15 mm. diameter the whorl height rapidly increases, the flank and the venter gradually flatten, and the whorl section develops a polygonal (high octagonal) outline. At first the primary ribs, and later also the secondaries, form a sharp angle at the peripheral margin as they pass on the venter.

Beginning with a diameter of 24–25 mm. the order of only one secondary rib between the primaries is established through a rapid shortening of the first and the third of the triple secondaries. The lateral tubercles on the primary ribs become weaker and more elongated, and eventually cease altogether. With the reduction and cessation of tubercles the primary ribs change from straight to flexed and are markedly attenuated at the midflank but retain their thickness at the umbilical edge, peripheral margin, and on the venter. In this process a few primary ribs may make an impression of trinodal ribs whereas actually no individualized bullae are formed. The secondary ribs are very thin below the midflank but with growth rapidly thicken toward the peripheral margin.

Between diameters of 40–45 and 62 mm., the angularity in the whorl section is reduced, the flanks are somewhat inflated, and the flat venter changes to subtabulate. The costation consists of alternating slightly flexed primary and secondary ribs.

The suture is not well perserved in any single specimen of the studied material. Enough has been learned, however, to show that it is typically acanthohoplitan, with a broad symmetrically trifid first lateral lobe and bifid saddles.

In the major characteristics the described species does not seem to differ greatly from *Paracanthohoplites multispinatus* (Anthula, 1899, p. 119, Pl. 10, figs. 5a–5c) which, however, retains more closely spaced ribs and tuberculation to a larger diameter, has two or one, instead of three, secondary ribs between the primaries in the corresponding earlier stages, and passes to the ornamentation of one secondary rib between two primaries in a less definite order. Anthula's reference to the arched peripheral area and subquadrate whorl section does not agree with the illustration of his holotype. It also is evident that there is only one tubercle on the primary ribs. There is, however, an appreciable depression in the ventral ribs in the early part of the last whorl, and apparently it was these thicker parts of the primary ribs at the umbilical edge and the peripheral margin that Anthula regarded as additional tubercles. The suture as drawn by Anthula, with a narrow first lateral lobe, seems to be too diagrammatic. A photographic reproduction of the suture in a variety of Anthula's species from Mangyshlak described by Sinzow (1908, p. 495, Pl. 7, figs. 1 and 2) as *Acanthohoplites multispinatus* var. *tenuicostata* shows a broad, symmetrically trifid first lateral lobe.

The assemblage of forms described by Sinzow (1908, p. 492, Pl. 7, figs. 1–8) as *A. multispinatus* (Anthula) is composed of different varieties, probably species, as is suggested by varietal names *"tenuicostata"* and *"robusta."* Some of these forms, like two specimens figured in his Figures 5 and 6 of Plate 7, may be close to Anthula's holotype. Kazansky (1914, p. 80) also believed that *Paracanthohoplites multispinatus* (Anthula) is represented in the material figured by Sinzow. Of interest is the presence of branching ribs at large diameters in some of Sinzow's examples, high on the flank as in his Figure 1, Plate 7, and in the umbilical half of the flank as in Figure 6, page 494. It should be noted in this connection that the penultimate whorl of Anthula's holotype has ribs that either branch or are very closely set at the umbilical edge.

In some of my paratypes the secondary ribs are closely spaced at the umbilical edge but do not branch. In one paratype there is a single branching primary rib in an early stage of growth. I also have a specimen, No. 91163 of my collection, not included in the paratypes, which at 4–4.5 mm. diameter shows two primary ribs that branch from lateral tubercles at the midflank, and also has depressed ribs on the peripheral area at 20 mm. diameter.

In my opinion, the suture and the presence of sporadic bifurcating ribs in certain trivial and varietal forms connect *Paracanthohoplites* with *Acanthohoplites*, whereas the siphonal depression of ribs in conjunction with a flat venter at comparatively large diameters is a characteristic observed in certain forms of the *Ammonites milletianus* group (d'Orbigny, 1840, Pl. 77, figs. 2 and 5).

Type: Holotype: No. 91150. Paratypes (illustrated): Nos. 91153, 91164, 91170; (not illustrated) 91144, 91152, 91161, 91162, 91965.

Occurrence: All types collected from limestone lenses in Baga shale, Joserita member, division 8e, of Lowell formation, Ninety One Hills.

Hypacanthohoplites Spath, 1923

Genotype: *Acanthoceras milletianum* (d'Orbigny) var. *plesiotypica* Fritel (1906, p. 245) = *Parahoplites jacobi* Collet (1907, p. 520, Pl. 8, fig. 1, only) by designation of Spath, 1923, p. 64.

Under the name *Hypacanthohoplites*[36] Spath (1923, p. 64) separated the group of *Ammonites mil-*

---

[36] The reason for emended spelling (orginally *"Hypacanthoplites"*) is the same as presented in the case of *"Acanthoplites."*

letianus d'Orbigny (1840, p. 263) as defined and restricted by Fritel (1906, p. 245) in the description of *"Acanthoceras" milletianum* (d'Orbigny) var. *plesiotypica*, a variety later advanced to the rank of species, *"Parahoplites" jacobi*, by Collet (1907, p. 520, Pl. 8, figs. 1-3). Fritel's only comment on his variety is that in some specimens the umbilical tubercles are conical and sufficiently well developed, whereas in others they are represented by bullae which grade into corresponding ribs. The smaller of his figures (p. 246, fig. 2) shows, as far as can be interpreted from a bibliofilm, both umbilical bullae and lateral tubercles with the primary ribs that branch from the latter. The variety *plesiotypica* is said to be sufficiently close to d'Orbigny's types. Collet's description of *"Parahoplites" jacobi*, based on three illustrated specimens including Spath's genolectotype, is essentially as below:

"Shell discoidal, whorls little involute and exposed 2/3 of their height. There are twenty-four ribs on the venter in one half of the whorl length at a diameter of 33 mm., and twenty ribs at a diameter of 70 mm., the largest diameter observed. Beginning with the whorl height of 10 mm. the flexuous primary ribs appear at the umbilical edge and cross the venter without any attenuation. In the younger stages, at the whorl height of 6 mm., the ribs are almost entirely attenuated [not developed] on the venter and are provided with small marginal tubercles [i.e., lateral tubercles placed close to the peripheral margin or the ventral bullae?]. The venter is flat up to the whorl height of 13 mm. After this whorl height is reached, the venter becomes slightly rounded. The straight secondary ribs appear at about the ventral third of the flank. In the young specimens, up to the whorl height of 6 mm., the peltoceratic primary ribs begin near the umbilicus where they form blunt tubercles [bullae?] and, after having formed a lateral tubercle, bifurcate toward the ventral portion of the flank. The primary ribs are straight until the point of bifurcation is reached, above this point they are slightly inclined forward." (Translated).

D'Orbigny did not discuss the presence of tubercles on the ribs in his description of *Ammonites milletianus* except to comment on the angular nature of the ribs at the peripheral margin. It may be inferred from the assemblage of prototypes in his monograph that the ribs are sparsely set in the young whorls; that at various stages the primary ribs are bullate either at the umbilical edge or at the midflank; that the bifurcation of the primary ribs, when present, takes place at the umbilical bullae; that the peripheral area is flat in adult stages; and that the ribs are attenuated or depressed on the venter of mature whorls.

The suture of *"Parahoplites" jacobi* drawn by Collet (1907, p. 521, fig. 2) is essentially of the same character as the suture figured by d'Orbigny (1840, Pl. 77, fig. 2) except that it is more symmetrical. The latter, however, was not taken from either of the two specimens illustrated in d'Orbigny's plate 77. It may be inferred from these two drawings that the first lateral lobe in the suture of *Ammonites milletianus* is trifid and symmetrical in attitude but with asymmetrically developed branches. The external and first lateral saddles are not as symmetrically bifid as in *Acanthohoplites*.

The essential difference between d'Orbigny's prototypes and Collet's *"Parahoplites" jacobi* is seen in (1) that the observable bifurcation of the primary ribs takes place only at the umbilical bullae in the former, whereas in the examples described and illustrated by Collet they also bifurcate higher on the flank from the lateral tubercles, and (2) that the venter of Collet's species loses its flat character, and the peripheral tubercles (ventral bullae?) cease, considerably earlier. Unfortunately Collet did not figure the peripheral view of Spath's genolectotype.

The following characteristics of *Hypacanthohoplites* may be of diagnostic value: In the younger stages the shell is with a flat peripheral area on which the ribs are weak, and on the flank the primary ribs bear only lateral tubercles placed at the peripheral margin. The umbilical bullae appear later with the passing of lateral tubercles to the flank. The branching of the primary ribs takes place at the midflank at lesser diameters but migrates toward the umbilical edge with growth. The development of ornamentation, therefore, is not essentially different from that observed in *Acanthohoplites*. If, however, the peripheral area and flanks are as flat in the earlier whorls as they are between the whorl height of 6 and 13 mm., *Hypacanthohoplites* probably does not develop a coronatiform shell in the early stage and in this respect may differ from *Acanthohoplites*.

In a more recent paper Spath (1939a, p. 236) reassigned the original genotype as *"Acanthohoplites" plesiotypicus* and selected the specimen illustrated by Collet (1907, Pl. 8, fig. 3) as the type of *"Acanthohoplites" jacobi*. The only ammonite discussed in that article as *Hypacanthohoplites* is

Ammonites milletianus d'Orbigny. I am not aware at the time of this writing whether there was a previous article redefining *Hypacanthohoplites* as a genus with a re-designation of the genotype.

Summarily, *Hypacanthohoplites* seems to stand for certain forms that definitely have a depressed ventral costation in the advanced stages of growth, as in the species and varieties of the *Ammonites milletianus* group, including those described by Collet (1907, p. 520, fig. 1; p. 522, fig. 3; p. 524, fig. 5; and p. 525, fig. 8) from the so-called *jacobi* zone of Vöhrum, Hanover. The genus appears to be valid but a re-study of the types is badly needed. The difference from *Acanthohoplites* may be accentuated even more if the presence of true ventral bullae can be shown to persist to the stage of full maturity in *Hypacanthohoplites*, as seems to be the case in *Hypacanthohoplites? milletianus* d'Orbigny (1840, Pl. 77, figs. 2 and 5).

Colombiceras Spath, 1923

GENOTYPE: *Ammonites crassicostatus* D'ORBIGNY (1840, p. 197, Pl. 59, figs. 1–2).

Spath (1923a, p. 64) placed his genus *Colombiceras*, proposed for the group of *Ammonites crassicostatus* d'Orbigny (1840, p. 197, Pl. 59, figs. 1–2), in the family Cheloniceratidae. In his designation of the genotype Spath mentioned neither d'Orbigny's younger specimen of *A. crassicostatus* (d'Orbigny, 1840, Pl. 59, figs. 4 and 4'), nor the suture drawn by d'Orbigny in Figure 3 of the same plate. There is no reason, however, to assume that this material does not belong in d'Orbigny's species and should not be included in the concept of *Colombiceras*, as already has been done "sub rosa" by Riedel (1937, p. 51), who also pointed out the asymmetry of cheloniceran first lateral lobe as compared with that in *Colombiceras*.

If the concept of Spath's genus is thus enlarged, the following features are to be considered. As in the majority of *Acanthohoplites* species, so in *Colombiceras* the branching of the primary ribs passes from lateral tubercles at the midflank to the umbilical edge during the early stage of growth. The suture, with bifid saddles and a symmetrically trifid first lateral lobe, also does not differ materially from that in *Acanthohoplites*. The distinguishing characteristic of the genus, therefore, is in the radial nature of the ribs, which are flat both on the flank and on the peripheral area.

It is doubtful that there is sufficient justification for the tendency of Spath (1921, p. 317, 318; 1923, p. 64, footnote 5) and Riedel (1937, p. 50) to connect with *Colombiceras* the flat-ribbed species for which the presence of tuberculate ribs has not been proved, as *Ammonites treffryanus* Karsten (1856, p. 109, Pl. 4, figs. 1a–1b) and *A. alexandrinus* d'Orbigny (1842, Pl. 2, figs. 8–11).

The development of forms with wedge-shaped ("en forma de cuña," Riedel, 1937, p. 50), flattened, or flat ribs in Aptian ammonites might have taken place independently in the Acanthohoplitinae, as has been shown by Sinzow (1908, p. 482, 484, Pl. 4, figs. 3–4; Pl. 5, figs. 9–13 and 16) in *Acanthohoplites laticostatus* and *A. subpeltoceroides*, and, less probably, either in Parahoplitinae, as suggested by Spath (1930, p. 440–441), who regarded *"Acanthohoplites" tobleri* (Jacob) as a transitional form between *Parahoplites* and *Colombiceras*, or even in the Deshayesitinae through the loss of the early tuberculation and ventral interruption of the ribs. A derivation from Parahoplitinae does not seem plausible since the untuberculate *"Colombiceras"* species have a different ornamentation and do not show the region of "crowded" ribbing in the early whorls, which is so characteristic of *Parahoplites*. Quite the opposite, as has been pointed out by Sinzow (1908, p. 485), in *Ammonites treffryanus* Karsten the ribs are distantly spaced in the penultimate whorl, in marked contrast with the strongly sigmoidal dense costation of the last whorl. Sinzow believed that this species belonged to an unknown genus. Such species, even if they have flattened ribs, are too remote from *Colombiceras crassicostatum*, because of their flexed costation and lack of tuberculation, to be congeneric with d'Orbigny's species. Unfortunately, the inner whorls of such forms have not been adequately studied, and there is no certainty at present that they do not possess tuberculate ribs in the early stages of development.

Therefore, in accepting *Colombiceras* in Spath's and Riedel's interpretation, it is expedient, at least for the present, to discriminate between two groups which may or may not be related: (1) the *Colombiceras crassicostatum* group, as has been proposed by Spath, and (2) the *Colombiceras? treffryanum* group. A third group, represented by rursicostate forms with flat ribs near the peripheral area and on the venter, contains too widely separated and too little known forms, such as *"Parahop-*

lites" treffryanus Anthula (1899, p. 115, Pl. 8, figs. 6a–6d, non *A. treffryanus* Karsten, 1856) with tuberculate ribs in the early whorls, or *"Parahoplites" umbilicostatus* Scott (1940, p. 1029, Pl. 62, fig. 8; Pl. 63, fig. 10) with an angular peripheral margin and closely set branching ribs in the mature stage, to allow an adequate conception as to their relationship.

Colombiceras? brumale Stoyanow, sp. nov.

(Plate 21, figures 8–10)

MEASUREMENTS:

	Holotype No. 91693
Diameter	46
Greater radius	28
Lesser radius	18
Height of last whorl	20
Thickness of last whorl	18
Width of umbilicus	15

Development of the early whorls of this species is unknown. The penultimate whorl has flat flanks and subtabulate peripheral area. Early in the last whorl the peripheral margin is less differentiated, and very rapidly with growth the whorl acquires a nearly circular section, moderately inflated and slightly converging flanks, a rounded venter, and a high umbilical wall. In the penultimate whorl the ribs are straight and distantly spaced; whether they are branching and tuberculate in the upper third of the flank, covered by the last whorl, is not known. The secondary ribs appear high on the flank in the last part of the penultimate and the early part of the last whorls. At first there is only one secondary rib between each pair of the primaries. At the whorl height of 10 mm., two secondaries appear and pass farther down the flank. At the whorl height of 12 mm., there are three longer secondary ribs, of which the first one nearly joins the adjacent primary rib at the umbilical edge. Then there are twice again only two secondary ribs between the primaries to the whorl height of 14 mm., in each of these pairs the anterior rib is the longer. In this part of the shell the primary ribs are slightly flexed on the flank and thickened at the umbilical edge. All the ribs are stronger on the peripheral area. At the above indicated whorl height there is a single primary rib which bifurcates at the midflank.

In the later, testiferous, half of the last whorl the closely set primary and secondary ribs alternate regularly. All the ribs are flattened in the upper part of the flank and rounded on the venter. The sigmoidal primary ribs are thin in the steep umbilical wall but rapidly increase in thickness as they pass over the umbilical edge. In the umbilical half of the flank the secondary ribs are represented by thin striae which at the midflank grade into a rib, very thin at first but equalling a neighboring primary rib in the peripheral half of the flank. There are also striae between the ribs. The suture is acanthohoplitan, with a symmetrically trifid first lateral lobe and bifid saddles.

In a general way this species resembles *Colombiceras? treffryanum* (Karsten, 1856, p. 109, Pl. 4, figs. 1a–b), a species with a greater whorl height, which also differs from *C.? brumale* in the development of costation. The ribs early acquire a subsigmoidal character. In the earlier part of the last whorl the primary ribs trifurcate low on the flank and, with further growth, are superseded by bifurcating primary ribs which irregularly alternate with single primaries and secondaries. In the terminal half of the last whorl the sigmoidal primary ribs uniformly alternate with single secondaries, as in the described species, but are less densely spaced.

Incomplete specimens and fragments with flat or flattish ribs are not infrequent in the stratum from which *C.? brumale* was collected. One of such fragments is shown in Figure 5, Plate 25. An incomplete specimen, No. 91699 of my assemblage, collected from the stratigraphically higher division 3b, differs from the form illustrated by Sinzow (1908, Pl. 5, figs. 11–12) as *Acanthohoplites laticostatus* in a lesser whorl height and a less angular peripheral margin.

TYPE: Holotype: No. 91693.

OCCURRENCE: Holotype collected from Quajote member, division 3d, of Lowell formation, Ninety One Hills.

Subfamily DESHAYESITINAE Stoyanow, subfamily nov.

This subfamily is proposed for *Deshayesites* Kazansky (1914, p. 99) and *Dufrenoya* Burckhardt (*in* Kilian, 1915, p. 34),[37] two genera distinguished from the Parahoplitinae by the following characters: (1) The costation gradually changes from dense in the inner whorls to more distantly spaced in maturity without a region of "crowded" ribs. The ribs are essentially sigmoidal on the flank and in the early adult stage are interrupted on the peripheral area (d'Orbigny, 1840, Pl. 85, figs. 1–4;[38] Leymerie, 1842, Pl. 17, figs. 17a–17b; Forbes, 1845, Pl. 5, fig. 2); (2) In the suture the external saddle considerably disagrees with the first lateral saddle in width, outline, and lobation. The first lateral lobe is subsymmetrically tripartite (in *Deshayesites deshayesi*, d'Orbigny, 1840, Pl. 85, fig. 4, and Forbes, 1845, Pl. 13, fig. 2; in *D. consobrinus*, Spath, 1930, p. 438, fig. a; and in *Dufrenoya justinae*, Scott, 1940, p. 1024; fig. 155; but not in *Deshayesites consobrinus*, d'Orbigny, 1840, Pl. 47, fig. 3, and *Dufrenoya dufrenoyi*, d'Orbigny, 1840, Pl. 33, fig. 6).

The Deshayesitinae differ sufficiently from the Acanthohoplitinae, both in the costation and the tuberculation, to permit a ready separation. The representatives of the former subfamily do not possess tubercles on the flank that notably change their relative position with growth of the shell.

Kazansky's (1914, p. 99) diagnosis of *Deshayesites* reads as follows:

"Shell flat, discoidal, with high whorl section and with peripheral area flattened in the young stage but more rounded in the maturity. Ribs more or less sickle-shaped, primaries beginning at the umbilical edge, secondaries at the midflank and equalling the primary ribs in thickness toward the periphery. In the young whorls all ribs cease before reaching the flattened venter. In the mature whorls all ribs cross the peripheral area with an attenuation at first, later without it, and eventually become thicker on the rounded periphery than on the flanks. In the suture, the external lobe is shorter than the first lateral lobe, sometimes considerably so, external saddle bipartite with a more developed internal branch, first lateral lobe appreciably broad and asymmetrical, first lateral saddle long and narrow, second lateral lobe about half the size of the first lateral lobe, second lateral saddle bifid and followed by small auxiliary lobe and saddle." (Translated).

Besides the groups represented by *Deshàyesites deshayesi* (Leymerie) and *D. weissi* (Neumayr and Uhlig), Kazansky also placed the *Dufrenoya dufrenoyi* (d'Orbigny)[39] group in *Deshayesites*.

According to Kazansky's interpretation, *Dufrenoya* is close to *Neocomites:* The ornamentation in the young whorls is almost identical, and in the first genus the characteristic thickening of the ribs at the peripheral margin begins to appear not earlier than at 10–15 diameter. *Deshayesites* is distinguished from *Neocomites* by the rounded and costate periphery in the mature whorls. Kazansky emphasized the resemblance between *Deshayesites* and *Dufrenoya* in the young stages, though in the latter genus the peripheral area remains flat in the mature whorls, thus approximating *Neocomites*.

He pointed out, further, that in *Dufrenoya* the peripheral interruption of the ribs ceases before 20 mm. diameter is reached, and at 20–25 mm. diameter it perceptibly differs from *Deshayesites* in the ribs being thicker at the peripheral margin. The two genera are again somewhat similar at 40–45 mm. diameter where the ribs cross the venter. At this stage, however, *Dufrenoya* has a flatter periphery and thicker ribs. In advanced growth stage of *Dufrenoya* Kazansky noted the presence of a forward arching of the ribs on the still flat peripheral area.

There is certain confusion in the nomenclature of discussed genera. Through an oversight, Roman (1938, p. 347) gave preference to *Parahoplitoides* Spath (1922, p. 111) over *Deshayesites* Kazansky (1914, p. 99), and to *Stenhoplites* Spath (1922, p. 110) over *Dufrenoya* Burckhardt (*in* Kilian, 1915, p. 34). The necessary corrections had already been made by Spath (1930, p. 424, 435).

In placing the Deshayesitinae in the same family with the Parahoplitinae, that is under the Para-

---

[37] Originally *Deshayesites* was introduced by Kazansky as a subgenus of *Hoplites*, and *Dufrenoya* was understood by Kilian as a subgenus of *Parahoplites*. Note Kilian's spelling: *Dufrenoyia* in p. 34, 35, 37, but *Dufrenoya* in p. 199 and 205. A generic disgnosis of *Dufrenoya* was given at a later date by Burckhardt (1925, p. 15).

[38] Spath (1930, p. 424) does not include in *Deshayesites deshayesi* the smaller specimen illustrated by d'Orbigny (1840, Pl. 85, fig. 3)

[39] Kazansky (1914, p. 106) identified this species with *Dufrenoya furcata* (J. Sowerby). Spath (1930, p. 435) has shown that the two species are not identical. Sowerby's species is rare; its holotype is an imperfectly preserved body-chamber cast.

hoplitidae Spath, I followed Spath and partly Roman. The development of ornamentation, as suggested by Kazansky, but especially the nature of suture, indicates a relationship with the Neocomitinae. However, the first lateral saddle in the Deshayesitinae is relatively broader and the following lobes and saddles are not inclined or oblique (Sarasin, 1897, p. 768, fig. 4; p. 769, fig. 6), approximating in this respect the simpler suture of *Parahoplites*.

Dufrenoya Burckhardt, 1915

GENOTYPE: *Ammonites dufrenoyi* D'ORBIGNY (1840, p. 200, Pl. 33, figs. 3–6).

In the specimens of *Dufrenoya dufrenoyi* of 17–20 mm. diameter, that I have from Vaucluse, France, the venter is narrowly rounded and crossed by the ribs without a peripheral angulation or tuberculation to the whorl height of 3–5 mm. Above this whorl height the ventral ribs are interrupted on the siphonal line and acquire a straight character through a rapid thickening at the peripheral angle. Gradually the ribs on the venter recede toward the peripheral margin, and at the whorl height of 8 mm. the venter is flat and ribless. The suture has a subsymmetrically tripartite first lateral lobe. In the suture of the holotype the lobes are defective according to d'Orbigny (1840, p. 202, Pl. 33, fig. 6), who represented the suture with a very asymmetrically bipartite first lateral lobe, not in agreement with his description. Probably the suture drawn by Sarasin (1897, p. 769, fig. 6) is typical for the genus.

The material referable to *Dufrenoya* collected from the Lowell formation is in fragmentary condition. Forms apparently conspecific with *Dufrenoya justinae* (Hill) are found in the Cholla member, division 5g, and a new species, described as *Dufrenoya joserita*, belongs in the Joserita member, division 8d. Besides, the Cholla member contains a great number of specimens probably congeneric with *Dufrenoya*. The material is represented only by the parts of larger whorls. This condition of preservation results from a previous selective mineralization and a subsequent disintegration of the inner whorls.

The difficulty in the interpretation of this material is the abundance in the same beds of large specimens which, though possessing flat flanks and costation of *Dufrenoya*, develop a rather arched peripheral area, and yet seemingly are connected by transitional forms with examples of *Dufrenoya* type (Pl. 22, figs. 1–8; Pl. 23, figs. 1–3). It was suggested that these incomplete specimens may belong to large forms of *Dufrenoya*. If this is so, they certainly are not conspecific with *D. justinae* (Hill). Hill's species has a high narrow whorl section, and flanks and periphery which remain flat at least to a diameter of 110 mm. (Burckhardt, 1925, Pl. 10, figs. 14–15). It is true, that in certain less compressed species, like *Dufrenoya truncata* Spath (1930, p. 436, Pl. 16, figs. 4a–4c), the venter seems to become somewhat arched, or at least less flattened with age, but the greatest resemblance to *Deshayesites* in such species is in the earlier and not in the later whorls.

A few fragments with somewhat eroded sutures were found. The better preserved suture of the larger illustrated specimen (Pl. 23, fig. 3) shows bifid saddles and an asymmetrically tripartite, nearly bipartite, first lateral lobe with more developed inner branches. The sutures of the smaller specimen (Pl. 22, fig. 3) have first lateral lobes with a stronger tripartition. Other specimens show that the tripartition of the first lateral lobe is less asymmetrical in the younger whorls. The first lateral lobe of this species is somewhat similar to that of *Deshayesites consobrinus* (d'Orbigny, 1840, Pl. 47, fig. 3), in which the inner branches of the lobe likewise are more strongly developed. The described suture is altogether unlike that of *Dufrenoya justinae* (Hill) illustrated by Scott (1940, p. 1024, fig. 155) which, by the way, is almost an exact replica of the suture of *Deshayesites consobrinus* (d'Orbigny) from England, as figured by Spath (1930, p. 438, fig. a).

Dufrenoya justinae (Hill)

(Plate 21, figures 11–17)

1893. *Acanthoceras* (?) *justinae* HILL, Proc. Biol. Soc. Washington, vol. 8, p. 38, Pl. 7, figs. 1–3.
1901. *Ammonites justinae* HILL, U. S. Geol. Survey, 21st Ann. Report, pt. 7, Pl. 21, fig. 6.
1925. *Dufrenoya justinae* BURCKHARDT, Inst. Geol. Mexico, Bol. 45, p. 17, Pl. 10, figs. 14–15.
1940. *Dufrenoya justinae* SCOTT, University of Texas Publication 3945, p. 1022, Pl. 60, figs. 7–8; Pl. 62, fig. 9.[40]

[40] For complete synonomy see Scott(1940, p.1022).

The illustrated fragments referred to Hill's species appear to be fairly close in whorl shape and ribbing to the casts of Hill's holotype and a paratype illustrated by Burckhardt (1925, Pl. 10, figs. 14-15) in my possession. In plesiotype No. 91653, of the whorl height between 21 and 25 mm., the flexuosity of ribs probably is relatively stronger, more as in Scott's densicostate plesiotype (Scott, 1940, Pl. 60, figs. 7-8). Specimens with more converging flanks, as in the latter form from Texas, also are represented in my assemblage, their ribs, however, are more distantly spaced. Fragment No. 91526, of 25 mm. whorl height, is identical with the corresponding part of Hill's holotype.

TYPE: Plesiotypes: Nos. 91653 and 91526.

OCCURRENCE: Both plesiotypes collected from the Cholla member, division 5g, of Lowell formation, Ninety One Hills.

Dufrenoya joserita Stoyanow, sp. nov.
(Plate 23, figures 4-6)

The shell of this species is known only in separated fragments. The smaller cotype, No. 91169, is a part of the whorl with a flat venter and with flanks which are flat and parallel to the whorl height of 11 mm. but slightly converge before the whorl height of 13 mm. is reached. The closely set ribs are considerably thicker at the peripheral margin and cross the venter without interruption. The subsigmoidal primary ribs bifurcate, and the secondary ribs appear, low on the flank. With growth of the shell the flexuosity of the more widely spaced ribs decreases. In the larger syntype, No. 91142, the parallel flanks change to slightly inflated between the whorl height of 22 mm. and 30 mm., and simultaneously the flattened venter changes to more rounded.

D. justinae (Hill) retains flat or slightly convergent flanks to a more advanced stage of growth. In *D. subfurcata* (Kazansky, 1914, p. 108, Pl. 7, fig. 92) the venter passes from flat to gently arched nearly as in the described species, but the secondary ribs are placed higher on the flank.

TYPE: Syntypes: Nos. 91142 and 91169.

OCCURRENCE: Both syntypes collected from Joserita member, division 8d, of Lowell formation, Ninety One Hills.

Dufrenoya? compitalis Stoyanow, sp. nov.
(Plate 22, figures 1-8; Plate 23, figures 1-3)

This species is based on fragmentary material collected in close association from a stratum 3 feet thick in the Cholla member. As far as can be interpreted from a large assemblage of conspecific separate whorls between 20 and 55 mm. in whorl height, the whorl section changes with growth from narrow oval, with a rather narrow subtabulate venter and flat flanks, which slightly converge in their peripheral third, to broadly oval. The venter, however, is always flattened, independently of the whorl section, and in larger whorls with a better differentiated peripheral margin is quite broad. The ribbing is coarse. The primary ribs vary from subsigmoidal to flexuous and change little in thickness between the umbilical edge and the peripheral margin. The bifurcating point of the primary ribs and the secondary ribs are placed high on the flank. All the ribs are straight and somewhat stronger on the venter.

The costation probably resembles more that in the coarse-ribbed species of *Deshayesites*, but the less sigmoidal ribs lack the characteristic forward inclination at the peripheral margin, and the younger whorls undoubtedly have more converging flanks and flatter venters.

The suture of two examples of this assemblage has been discussed under *Dufrenoya*.

TYPE: Syntypes: Nos. 91645, 91661, 91547, 91646, 91683, 91662.

OCCURRENCE: All syntypes were collected from Cholla member, division 5g, of Lowell formation, Ninety One Hills.

Deshayesites Kazansky, 1914

GENOTYPE: *Ammonites deshayesi* LEYMERIE (in d'Orbigny, 1840, p. 288, Pl. 85, figs. 1-4).

In the genotype the early ribs are sickle-shaped, with a shorter forward convexity in the umbilical part of the flank and a longer sinuosity in its upper part. With growth of the shell this ribbing gradually changes to sigmoidal. Sickle-shaped ribs, however, are not in evidence in the illustrations of Leymerie (1842, Pl. 17, fig. 17) and Forbes (1845, Pl. 5, fig. 2). In the species represented by larger forms the ribs are essentially sigmoidal (d'Orbigny, 1840, Pl. 47, fig. 1; Neumayr and Uhlig, 1881, Pl. 46, fig. 1; Koenen, 1902, Pl. 9, figs. 1a and 2a). In the majority of species of this genus the forward bending of the ribs on the flank near the peripheral margin is retained to larger diameters of the shell. In some species, however, as in *Deshayesites grandis* Spath (1930, Pl. 17, fig. 2a), the sigmoidal ribbing is gradually replaced by flexuous ribs that are nearly straight at the peripheral margin between diameters of 90 and 110 mm.

Two described species from the Lowell formation are placed provisionally in *Deshayesites*, although they can be distinguished at once from the typical species of Kazansky's genus by their less sinuous costation, the low position on the flank of the bifurcating point of the ribs, and the lack of a distinct forward inclination of the ribs near the peripheral margin at larger diameters. Nevertheless, they have certain features in common with species which, though deviating from the genotype, have been associated with the genus.

Deshayesites? temerarius Stoyanow, sp. nov.

(Plate 24, figures 1–3)

This species is represented by an incomplete and somewhat deformed specimen of 140 mm. diameter with an oval whorl section, slightly convex flanks, and a narrowly rounded peripheral area. The whorl thickness is 11 mm. at 29 mm. whorl height, between the whorl height of 43 mm. and 64 mm. it increases from 19 mm. to 32 mm. The primary ribs are strong in the umbilical wall and gain in thickness very little above the umbilical third of the flank. These ribs are subsigmoidal to 50 mm. whorl height but become flexuous and even straight in the terminal part of the final whorl where they are somewhat attenuated at the midflank. The bifurcating point of the primary ribs and the lower limit of the secondary ribs are slightly below the midflank in the earlier stages but gradually migrate toward the umbilical edge in the last whorl. All the ribs are straight on the venter which they cross without attenuation. The suture is not satisfactorily enough preserved for an adequate interpretation except that the lobules are well rounded and leaf-like.

In the whorl section and the peripheral area this form resembles the species with a narrow oval cross section represented by *Deshayesites consobrinus* (d'Orbigny, 1840, Pl. 47, fig. 2). The costation appears to be more comparable to certain forms of the *D. weissi* group that have a tendency toward attenuation of the ribs at the midflank of larger whorls (Neumayr and Uhlig, 1881, p. 51). Apparently the periphery in such forms varies from rounded or narrowly arched to subtabulate (Spath, 1930, p. 426–429, Pl. 17, fig. 2b). In this group, however, the sinuosity of more sigmoidal ribs is stronger, and at larger diameters both the secondary ribs and the branching point of the bi- or even tri-furcating primary ribs are higher on the flank.

TYPE: Holotype: No. 91222.

OCCURRENCE: Holotype collected from thin-bedded limestone above a grit tentatively correlated with Espinal grit in a ridge 1½ miles northeast of Black Knob Hill.

Deshayesites? butleri Stoyanow, sp. nov.

(Plate 25, figures 1–3)

The large specimen on which this species is based was found in the Espinal grit. Unfortunately removed from the grit not carefully enough by too ardent collectors, it is represented by disconnected whorl parts of 22 mm., 36 mm., 60 mm., and 86 mm. whorl height. In the observed sections the whorl outline changes from narrow oval to oval with almost equally curved umbilical edge and peripheral margin, and with high flat flanks. At 22 mm. whorl height the thickness of the whorl is 10 mm., and at 86 mm., taken at the midflank, it is 42 mm. The peripheral area passes with growth from the narrowly rounded, through rounded, to broadly subtabulate. In the earliest observed whorls, of the specimen (not illustrated) of 22 mm. whorl height, the undifferentiated single ribs are distantly spaced and straight from the umbilical edge to the peripheral area. Whether they cross the latter at this stage is not known. In the larger whorls the ribs are strong in the umbilical wall, attenuated and even smoothed about the midflank, and distinctly stronger again in the peripheral part of the flank and on the venter. In the last observed whorl they change from somewhat flexuous to nearly straight, bifurcate slightly below or at the midflank, and do not show a forward inclination at the peripheral margin. The suture is not preserved.

The ornamentation of *D.? butleri* is of the same general type as in *D.? temerarius*, but the ribs are weaker on the flank, relatively stronger at the umbilical edge, and more attenuated at the midflank. However, this species, with its straight ribs in the earlier whorls and a broad periphery in the mature stage, seems to be more remote from the forms of the *D. weissi* group that show a tendency to lateral smoothness in the larger whorls. The low bifurcating, nearly straight and attenuated on the flank, ribs of *D.? butleri* more resemble the lateral costation of older species, like "*Hoplites*" *scioptychus* Uhlig (1901, p. 57, Pl. 5, fig. 1a).

TYPE: Holotype: No. 91204.

OCCURRENCE: Holotype collected from Espinal grit, Joserita member, division 8c, of Lowell formation, Ninety One Hills.

Family DESMOCERATIDAE Zittel, emend. H. Douvillé, 1916

Beudanticeras Hitzell, 1915

Beudanticeras victoris Stoyanow, sp. nov.

(Plate 18, figures 18–21)

MEASUREMENTS:

	Holotype No. 91909
Diameter	55
Greater radius	30
Lesser radius	25
Height of last whorl	24
Thickness of last whorl	18
Width of umbilicus	12

The partly testiferous holotype of this species is discoidal, with the involution changing from two thirds to one half of the whorl between diameters of 10 mm. and 55 mm. The early shell has a rounded venter and nearly parallel, slightly inflated flanks to the whorl height of 5 mm. At about 15 mm. whorl height the flanks converge toward the narrowly rounded peripheral area. In the last whorl the venter is gently arched; the whorl section is suboval, wider in its umbilical part and more compressed between the peripheral parts of the flanks in which the outer spiral depressions of the shell are distinctly indicated up to 45 mm. diameter; the inner spiral depressions, however, are barely perceptible on the flanks and do not show well in the whorl section. The narrow umbilicus has steeply inclined but not vertical walls. The umbilical edge is not sharply defined.

The ornamentation of inner whorls consists of closely set sigmoidal bulges and constrictions of varying prominence, as strongly developed on the peripheral area as on the flank, some of them, however, are stronger in the umbilical half of the flank, others, in its peripheral half and on the venter. The last whorl has seven wider-spaced sigmoidal bulges and as many constrictions, all of which cross the venter. The preserved test shows fine, closely set falciform striae. When viewed in a strong reflected light the test suggests that besides the bulges and the striae there might have been very weak ribs on the flank.

The suture is partly preserved only in the inner whorls. At a diameter of 10 mm. the narrow external lobe is shorter than the first lateral lobe. The external saddle is almost symmetrically bifid and has nearly parallel sides. In the asymmetrically trifid first lateral lobe a larger outer branch, which does not undercut the external saddle, and a narrow tripartite terminal branch are distinguished.

Beudanticeras victoris is close to *Beud. walleranti* (Jacob) = "*Desmoceras (Uhligella)*" *walleranti* Jacob (1907, p. 31, Pl. 3, figs. 1a–1c, only) from which it differs essentially, as well as from *B. stolic-*

zkai (Kossmat, 1898. *Compare* Spath, 1921, p. 275-276), in the presence of strongly developed bulges and constrictions in the inner whorls.

TYPE: Holotype: No. 91909.

OCCURRENCE: Holotype collected from Quajote member, division 3d, of Lowell formation, Ninety One Hills.

Family LYELLICERATIDAE Spath, 1931.

Stoliczkaia Neumayr, 1875

GENOTYPE: *Ammonites dispar* D'ORBIGNY (1840, p. 142, Pl. 45, figs. 1-2).

Specimens of *Stoliczkaia* referable to several species are abundant in the Molly Gibson formation of the Patagonia Mountains, south of Tucson. The preservation is not very good, however, and in my collection of over 100 better preserved examples not a single one shows the suture. This condition is due to the nature of cementation by a rather soft arenaceous limestone in which the grains of sand adhere to calcareous shells.

Although the scaphitoid habit is observed in certain of these forms, especially in shells of smaller diameters, all of them differ from the genotype in retaining a more or less strongly developed costation in the mature whorls. The material is conveniently separated into: 1. Forms in which coarse, blunt ribs, replacing the denser and finer costation of earlier whorls, appear at comparatively smaller diameters, about 40 mm. on the average, and which approximate the varieties of *S. notha* (Seeley, *in* Spath, 1931, p. 335); 2. More inflated forms with a pronounced forward sweep of the ventral ribs, which probably are closer to the varieties of *S. dorsetensis* Spath (1931, p. 337); 3. Apparently connected with these but appreciably different are densicostate forms with a broad flat venter and gently angular peripheral margin; 4. Coarsely ribbed forms that remain narrowly compressed and densicostate to larger diameters, lack a forward sinus in the ribs on the venter, and often show a blunt angulation at the peripheral margin. Variability is considerable and there are intermediate forms.

Discussion and illustration of the representative forms of these groups is beyond the scope of this paper and will be presented elsewhere. Three species, distinguished by a more pronounced differentiation of the primary and secondary ribs than is observed in the majority of British examples, are described below.

Stoliczkaia patagonica Stoyanow, sp. nov.

(Plate 26, figures 3-4)

MEASUREMENTS:

	Holotype No. PT23	Paratype No. PT32	Paratype No. PT51
Diameter	53	66	26
Greater radius	33	39	15
Lesser radius	20	27	11
Height of last whorl	24	31	12
Thickness of last whorl	12	18	10
Width of umbilicus	17 ca.	20 ca.	Not obs.

A coarse-ribbed species, strongly compressed to the largest observed diameter of 66 mm. The flat flanks meet the flattened or slightly convex peripheral area at a right angle, but the peripheral margin is rounded. A relative increase in the whorl thickness is not appreciable until a diameter of 60 mm. is reached. At the earlier stages of growth the ribs are little differentiated and the secondary ribs are concentric with the primaries. Beginning with the whorl height of 19-20 mm. the distinctly stronger primary ribs are flexed, and the shorter secondary ribs, predominantly two between the primaries, are oriented at an angle to the latter. All the ribs are straight and stronger on the venter.

The species from the Main Street and Grayson formations of Texas described by Scott (1926, p. 141, Pl. 3, figs. 3-4) as *Stoliczkaia dispar* (d'Orbigny) is somewhat similar in a lateral view. It essentially differs from the present species in its rounded venter and rapid increase in the whorl thickness, and may be close to *S. dispar* as has been pointed out by Scott and Spath (1931, p. 337).

Type: Holotype: No. PT23.   Paratypes (not illustrated): Nos. PT32 and PT51.

Occurrence: Arenaceous limestone in upper part of Molly Gibson formation, west of Molly Gibson mine, Patagonia Mountains.

Stoliczkaia excentrumbilicata Stoyanow, sp. nov.

(Plate 26, figures 5–6)

Measurements:

	Holotype No. PT01
Diameter	24
Greater radius	14
Lesser radius	8
Height of last whorl	14
Thickness of last whorl	11
Width of umbilicus	12

In this species the scaphitoid habit is observed in shells of small diameters, which probably represent the full-grown stage. In the holotype, the excentrumbilication is quite pronounced, as also is the rapid increase in the height and thickness of the last whorl. On the flank the longer primary ribs with a general forward inclination are strongly indicated. In the last whorl there are two secondary ribs between each pair of the primaries placed slightly at an angle to the latter, except in the terminal part of the whorl where the primary and secondary ribs alternate and are concentric. On the venter the forward sweep of the ribs is well developed and the ribs appreciably broaden with the growth of the shell. The costation of this species differs from that of *Stol. patagonica* in the less flexed primary ribs and the presence of a peripheral sweep.

Type: Holotype: No. PTO1.

Occurrence: Arenaceous limestone in upper part of Molly Gibson formation, west of Molly Gibson mine, Patagonia Mountains.

Stoliczkaia scotti Stoyanow, sp. nov.

(Plate 26, figures 7–8)

Measurements:

	Holotype No. PT10
Diameter	45
Greater radius	27
Lesser radius	18
Height of last whorl	18
Thickness of last whorl	17 ca.
Width of umbilicus	15 ca.

This species, with flexed primary ribs in the last whorl that are bullate in the umbilical part of the flank, and with three shorter concentric secondary ribs between the primaries, resembles *Stoliczkaia dorsetensis* Spath (1923, Pl. 32, fig. 7) from the *dispar* zone of White Nothe, England, from which it differs in flatter flanks, although the latter are not as flat as it appears in my illustration. The ribs on the venter do not show a forward sweep and flattening until the terminal part of the last whorl is reached, where they are as much bent forward as in *S. dorsetensis* (Spath, 1931, Pl. 31, figs. 10, 14, and 2) but are not as much broadened.

Type: Holotype: No. PT10.

Occurrence: Arenaceous limestone in upper part of Molly Gibson formation, west of Molly Gibson mine, Patagonia Mountains.

INCERTAE SEDIS

Among imperfectly preserved specimens collected in the studied area there are two that should be recorded for further research.

Cleoniceras? schlaudti Stoyanow, sp. nov.
(Plate 24, figures 4-5)

This weathered specimen, of 155 mm. diameter, was found loose in the dry wash southeast of Schlaudt Ridge (*see* Plate 27). Its matrix is suggestive of the limestones in the lower part of the sequence in the Ninety One Hills, like the limestone in division 9c of Pacheta member. The anterior part of the last whorl is worn away thus partly exposing the penultimate whorl. In the flank of the latter are seen slightly inclined forward and closely set ribs which apparently extend to the venter. In the exposed earlier part of the same whorl, at its contact with the present terminal edge of the final whorl, the ribs are better preserved. Here they are slightly flexed and nearly reach the umbilical edge. Unfortunately this is not satisfactorily seen in the illustration because of a thick layer of glue.

The described specimen somewhat resembles *Cleoniceras baylei* (Jacob, 1907, p. 59, Pl. 7, figs. 25a-25b) but has a much narrower, nearly acute, peripheral area at the corresponding whorl height. It is impossible to determine with certainty whether the irregularly radial depressions and elevations on the flank result from weathering or represent the bulges and constrictions similar to those of *Beudanticeras*.

SPECIMEN: No. 91068.

Douvilleiceras? muralense Stoyanow, sp. nov.
(Plate 26, figures 1-2)

The described specimen is a part of the penultimate whorl of a large ammonite found *in situ* in the assumed basal part of the Mural limestone, in Mural Hill Ridge northeast of Bisbee. The exposed last and inner whorls were completely fractured and badly eroded. The partly preserved fragment illustrated in Pl. 26, figs. 1-2, alone shows the costation and parts of the suture. In this fragment the height of the whorl above the venter is 95 mm. and the thickness is about 160 mm. The distantly spaced broad ribs are straight, coarser and stronger on the flank than on the venter which, however, may result from the nature of weathering. On the peripheral area, their anterior side is more steeply inclined, whereas posteriorly there is a gentle slope to the preceding rib. There is no evidence of tuberculation of the ribs or of their interruption on the venter. Of the sutures only the external lobes and external saddles are preserved.

By the character of the external saddle this specimen is not likely to belong in Parahoplitidae or Cheloniceratidae. Such external saddles, narrow and asymmetrically bifid in their terminal part, with a third strong branch on the outer side, have been observed in certain species of *Douvilleiceras* (*see* the copy of Quenstedt's figure of *D. inaequinodum* in Spath, 1925, p. 74, fig. 15a).

SPECIMEN: No. 94029.

OCCURRENCE: Gray ledge-forming limestone (24) in 4th escarpment of Mural Hill Ridge section regarded as base of Mural limestone in Mule Mountains.

I have pointed out elsewhere in this paper that Anthula (1899, Pl. 10, fig. 3b) represented the suture of *Acanthohoplites aschiltaensis* with a subsymmetrically bifid first lateral lobe, not in evidence in other species of this genus. In the course of research I tried to locate Anthula's types. This was not an easy task. His described material consisted of two sets: Abich's collection in University of Vienna, at present inaccessible, and Sjögren's collection deposited at University of Uppsala. It was in the latter collection, packed and removed to safety during the war, that Dr. H. G. Backlund, Professor in the University of Uppsala and Correspondent of the Geological Society of America, assisted by Dr. V. Jaanusson of the same institution, located one of Anthula's (1899, Pl. 10, figs. 2a, 2b) syntypes discussed in this paper, made excellent photographs of the suture, and sent them to me. As I had surmised, the suture is typically "acanthohoplitan," with a symmetrically trifid first lateral lobe and bifid saddles. It is too late to insert an illustration, this will be done in a separate article. I take this opportunity to express my deepest gratitude to Professor Backlund and Dr. Jaanusson for their valued cooperation.

REFERENCES

Adkins, W. S. (1928) *Handbook of Texas Cretaceous fossils*, Univ. Texas Bull. 2838, 385 p., 37 pl Austin.

────── (1932) *The Mesozoic System in Texas, in* Sellards, E. H., Adkins, W. S., and Plummer, F. B.: *The Geology of Texas*, vol. 1, *Stratigraphy, Part 2*, Univ. Texas Bull. 3232, p. 239–518, figs. 13–27. Austin.

────── and Winton, W. M. (1919) *Paleontological correlation of the Fredericksburg and Washita formations in North Texas*, Univ. Texas Bull. 1945, 128 p., 21 pls. Austin.

Agassiz, L. (1840) *Études critiques sur les mollusques fossiles*, Livraison 1, 58 p., 11 pls. Neuchâtel.

Albritton, Claude C., Jr. (1937) *Upper Jurassic and Lower Cretaceous ammonites of the Malone Mountains, Trans-Pecos Texas*, Harvard Coll., Mus. Comp. Zool. Bull., vol. 80, no. 10, p. 391–412, pls. 1–9. Cambridge, Mass.

────── (1938) *Stratigraphy and structure of the Malone Mountains, Texas*, Geol. Soc. Am., Bull., vol. 49, p. 1747–1806, pls. 1–9. New York.

Anderson, F. M. (1938) *Lower Cretaceous deposits in California and Oregon*, Geol. Soc. Am., Spec. Papers, no. 16, 339 p., 84 pls.

Anthula, Dim. J. (1899) *Ueber die Kreidefossilien des Kaukasus*, Beitr. Paläont. u. Geol. Österr. Ungarns u. des Orients, B. 12, p. 54–159, pls. 1–14. Wien.

Arkell, W. J. (1933) *The Jurassic System in Great Britain*, Oxford, Clarendon Press, 681 p., 41 pls.

Bigot, A. (1892) *Contributions à l'étude de la faune jurassique de Normandie: Mémoire sur les Trigonies*, Soc. Linnéenne de Normandie, Mém., vol. 17, ser. 2. Caen.

Bogdanowitch, Ch. (1890) *Notes sur la géologie de l'Asie centrale. 1. Description des quelques dépôts sedimentaires de la contrée Transcaspienne et d'une partie de la Perse septentrionale*, Russ.-K. Mineral. Gesell. St. Petersburg, Verhandl., vol. 26, p. 1–192, pls. 1–8.

Böse, E. (1910) *Monografía geológica y paleontológica del Cerro de Muleros cerca de Ciudad Juárez, Estado de Chihuahua, y descripción de la fauna cretácea de la Encantada, placer de Guadalupe, Estado de Chihuahua*, Inst. Geol. México, Bol., no. 25, 193 p., 48 pls. México, D. F.

────── and Cavins, O. A. (1927) *The Cretaceous and Tertiary of southern Texas and northern Mexico*, Univ. Texas Bull. 2748, p. 7–142. Austin.

Buch, L. von (1839) *Pétrifactions recueillies en Amérique par Mr. A. de Humboldt et par Charles Degenhardt*, Gesammelte Schriften, 4, p. 519–542, pls. 1–24. Berlin.

Burckhardt, Carl (1903) *Beitraege zur Kenntniss der Jura- und Kreideformation der Cordillere*, Palaeontographica, B. 50, 144 p., 16 pls. Stuttgart.

────── (1906) *La faune jurassique de Mazapil avec un Appendice sur les fossiles du Crétacique inférieur*, Inst. Geol. México, Bol., no. 23, p. 5–216, pls. 1–43. México, D. F.

────── (1912) *Faunes jurassiques et crétaciques de San Pedro del Gallo*, Inst. Geol. México, Bol., no. 29, 264 p., 46 pls. México, D. F.

────── (1925) *Faunas del Aptiano de Nazas (Durango)*, Inst. Geol. México Bol., no. 45, 71 p., 10 pls. México, D. F.

────── (1930) *Étude synthétique sur le Mésozoïque mexicain*, Soc. Paléont. Suisse, Mém., vols. 49–50, 280 p., 18 tables, 65 figs. Bâle.

Butler, B. S., Wilson, E. D., and Rasor, C. A. (1938) *Geology and ore deposits of the Tombstone district, Arizona*, Univ. Ariz., Ariz. Bur. Mines, Bull. no. 143, Geol. Ser. no. 10. Tucson.

Castillo, Antonio Del and Aguilera, Jose G. (1895) *Fauna fósil de la Sierra de Catorce, San Lui Potosí*, Com. Geol. México, Bol., no. 1. Mexico, D. F.

Chao, Y. T. (1928) *Productidae of China, Part 2*, Geol. Survey China, Palaeontologia Sinica, ser. B, vol. 5, fasc. 3, 103 p., 6 pls. Peking.

Collet, Léon W. (1907) *Sur quelques espèces de l'Albien Inférieur de Vöhrum (Hanovre)*, Soc. Phys. Hist. Nat.Genève, Mém., vol. 35, p. 519–529, figs. 1–10, pl. 8.

Conrad, T. A. (1855) *Description of one Tertiary and eight new Cretaceous fossils from Texas*, Acad. Nat. Sci. Philadelphia, Pr. 7, p. 268–269.

Conrad, T. A. (1857) *Descriptions of Cretaceous and Tertiary fossils. In* Emory, W. H., *Report of the United States and Mexican Boundary Survey*, U. S. 34th Cong., 1st Session, S. Ex. Doc. 108 and H. Ex. Doc. 135, vol. 1, pt. 2, p. 141–174, pls. 1–19 and pl. 21.

Coquand, H. (1865) *Monographie paléontologique de l'étage Aptien de l'Espagne*, Soc. d'Emulation de la Provence, Mém., vol. 3, p. 191–411, pls. 1–28. Marseille.

────── (1869) *Monographie du genre Ostrea, Terrain crétacé*, 215 p., 75 pls. Paris.

Cragin, F. W. (1893) *A contribution to the invertebrate paleontology of the Texas Cretaceous*, Geol. Survey Texas, 4th Ann. Rept., p. 139–246, pls. 24–46. Austin.

────── (1905) *Paleontology of the Malone Jurassic formation of Texas*, U. S. Geol. Survey, Bull. 266, p. 1–172, pls. 1–29. Washington.

Dana, James Dwight (1849) *Geology. United States exploring expedition, during the years 1838, 1839, 1840, 1841, 1842, under the command of Charles Wilkes, U. S. N.*, vol. 10, 756 p., pls.

Dane, Carle Hamilton (1929) *Upper Cretaceous formations of southwestern Arkansas*, Ark. Geol. Survey, Bull. 1, 215 p., 29 pls.

Darton, N. H. (1925) *A resumé of Arizona geology*, Univ. Ariz., Ariz. Bur. Mines, Bull. no. 119, Geol. Ser. no. 3, 298 p. Tucson.

────── (1928) *"Red Beds" and associated formations in New Mexico; with an outline of the geology of the State*, U. S. Geol. Survey, Bull. 794, 356 p. Washington.

Deussen, Alexander (1924) *Geology of the Coastal Plain of Texas west of Brazos River*, U. S. Geol. Survey, Prof. Paper 126, 139 p., 36 pls. Washington.

Dietrich, W. O. (1938?) *Lamelibranquios cretácicos de la Cordillera Oriental (Estudios geológicos y paleontológicos sobre la Cordillera Oriental de Colombia, parte tercera)*, Repúb. Colombia—Ministerio de Industrias y Trabajo, Dept. Minos y Petróles, p. 81–108, pls. 15–22.

Douvillé, H. (1904) *Mission scientifique en Perse par J. de Morgan*, t. 3, pt. 4, *Paléontologie, Mollusques fossiles*, 380 p., 25 (28?) pls., Paris.

Dumble, E. T. (1902) *Notes on the geology of southeastern Arizona*, Am. Inst. Min. Eng., Tr., vol. 31, p. 696–715. New York.

Forbes, E. (1845) *Catalogue of the Lower Greensand fossils in the Museum of the Geological Society with notices of species new to Britain*, Geol. Soc. London, Quart. Jour., vol. 1., p. 237–250; 345–355, pls.

Frech, Fritz (1911) *Abschliessende paleontologische Bearbeitung der Sammlungen F. von Richthofens*, *In:* Richthofen, Ferd. Fr. von, *"China,"* B. 5. Berlin.

Fritel, P. H. (1906) *Sur les variations morphologiques d*'Acanthoceras milletianum, *d'Orb. sp.*, Le Naturaliste, ser. 2, no. 472, p. 245–247. Paris.

Galliher, E. Wayne (1931) *Notes on excrement*, Micropaleont. Bull., vol. 3, no. 1, p. 11–12, 1 plate. Ann Arbor.

Gerhardt, K. (1897–1898) *Beitrag zur Kenntniss der Kreideformation in Columbien. In:* Steinmann, G. von, *Beiträge zur Geologie und Paleontologie von Südamerika*. Neues Jahrb. Min., Geol. V. Palaeont., Beil.-Bd. 11, p. 116–208, pls. 3–5. Stuttgart.

Gillet, S. (1922) *Étude du Barrêmien Supérieur de Wassy (Haute Marne)*. Soc. Géol. France, Bull., ser. 4, vol. 21, p. 3–47, pls. 1–3. Paris.

────── (1924) *Études sur les lamellibranches néocomiens*. Soc. Géol. France, Mém., ser. n., vol. 1, Mém. 3, p. 5–224, pls. 1–2. Paris.

Hill, R. T. (1893) *Paleontology of the Cretaceous formations in Texas: The invertebrate paleontology of the Trinity division*, Biol. Soc. Washington, Pr., vol. 8, p. 9–40, pls. 1–8.

Hyatt, Alpheus (1903) *Pseudoceratites of the Cretaceous*, edited by T. W. Stanton. U. S. Geol. Survey, Mon. 44, 351 p., 47 pls. Washington.

Imlay, Ralph W. (1939a) *Paleogeographic studies in northeastern Sonora*, Geol. Soc. Am., Bull., vol. 50, p. 1723–1744, pls. 1–4. New York.

────── (1939b) *Possible interoceanic connections across Mexico during the Jurassic and Cretaceous periods*, 6th Pacific Sci. Cong., Pr., p. 423–427.

────── (1944) *Correlation of the Cretaceous formations of the Greater Antilles, Central America, and Mexico*, Geol. Soc. Am., Bull., vol. 55, p. 1005–1046, pls. 1–3. New York.

Jacob, Charles (1905) *Étude sur les ammonites et sur l'horizon stratigraphique du gisement de Clansayes*, Soc. Géol. France, Bull., sér. 4, vol. 5, p. 399–432, pls. 12–13. Paris.

—— (1907) *Étude sur quelques ammonites de Crétacé moyen*, Soc. Géol. France, Mém., Paléont., Mém. 38, p. 5–63, pls. 1–9. Paris.

—— and **Tobler, Auguste** (1906) *Étude stratigraphique et paléontologique du Gault de la Vallée de la Engelberger AA*, Soc. Paléont. Suisse, Mém., vol. 33, p. 3–26, pls. 1–2. Genève.

Karsten, Hermann (1856) *Über die geognostischen Verhältnisse des westlichen Columbien, der heutigen Republiken Neu-Granada und Equador*, Amtlicher Ber. über die zwei und dreiszigste Versammlung deutscher Naturf. u. Aerzte zu Wien, p. 80–117, pls. 1–6. Vienna.

Kazansky, P. A. (1914) *Déscription d'une collection des Céphalopodes des terrains Crétacés du Daghestan*, Izv. Tomskago Technol. Inst., vol. 32, no. 4, 127 p., 7 pls. Tomsk.

Kellum, Lewis B. (1936a) *Paleogeography of parts of border province of Mexico adjacent to West Texas*, Am. Assoc. Petrol. Geol., Bull., vol. 20, no. 4, p. 417–432. Tulsa.

—— (1936b) *Evolution of the Coahuila Peninsula, Mexico: Part 3.—Geology of the mountains west of the Laguna District*, Geol. Soc. Am., Bull., vol. 47, p. 1039–1090, pls. 1–14. New York.

—— (1937) *Geology of the sedimentary rocks of the San Carlos Mountains. In: The Geology and Biology of the San Carlos Mountains, Tamaulipas, Mexico*, Univ. Mich. Stud., Sci. Ser., vol. 12, p. 3–97, pls. 1–9. Ann Arbor.

Kilian, W. (1913) *Unterkreide (Palaeocretacicum): Das bathyale Palaeocretacicum im südöstlichen Frankreich; Apt-Stufe, Urgonfacies im südöstlichen Frankreich, In:* Lethaea Geognostica, T. 2, B. 3, Abt. 1, Lief. 3, p. 289–398, pls. 9–14. Stuttgart.

—— (1915) *Contribution à l'étude des faunes paléocrétacées du Sud-Est de la France: 1. La faune de l'Aptien Inférieur des environs de Montélimar*, Mémoires pour servir à l'explication de la Carte Géologique Détaillée de la France, p. 3–221, pls. 1–9. Paris.

King, Robert E. (1939) *Geological reconnaissance in northern Sierra Madre Occidental of Mexico*, Geol. Soc. Am., Bull., vol. 50, p. 1625–1722, pls. 1–9. New York.

Kitchin, F. L. (1903) *The Jurassic fauna of Cutch. The Lamellibranchiata. Genus Trigonia*, Palaeont. Indica, ser. 9, vol. 3, part 2, no. 1, 122 p., 10 pls. Calcutta.

—— (1908) *The invertebrate fauna and palaeontological relations of the Uitenhage series*, South African Mus., Ann., vol. 7, pt. 2, no. 3, p. 21–250, pls. 2–11.

—— (1926) *The so-called Malone Jurassic formation in Texas*. Geol. Mag., vol. 63, p. 454–469. London.

Knight, J. Brookes (1941) *Paleozoic gastropod genotypes*, Geol. Soc. Am., Spec. Papers, no. 32, 509 p., 96 pls.

Kniker, Hedwig T. (1918) *Comanchean and Cretaceous Pectinidae of Texas*, Univ. Texas, Bull. 1817, 56 p., 10 pls. Austin.

Koenen, A. Von. (1902) *Die ammonitiden des norddeutschen Neocom (Valanginien, Hauterivien, Barrêmien und Aptien)*, K. Preuss. Geol. Landesanstalt u. Bergakad., Abh., Neue Folge, Heft 24, 451 p., 55 pls. Berlin.

Kossmat F. (1898) *Untersuchengen über die südindische Kreideformation*, Beitr. Paleont. u. Geologie Österr.-Ungarns u. des Orients, Bd. 11, p. 89–152, pls. 14–19. Wien.

Lasky, Samuel G. (1938) *Newly discovered section of Trinity age in southwestern New Mexico*, Am. Assoc. Petrol. Geol., Bull., vol. 22, no. 5, p. 524–540. Tulsa.

Leymerie, A. (1842) *Mémoire sur le terrain crétacé du department de l'Aube*, Soc. Géol. France, Mém., ser. 1, vol. 5, pt. 1, 34 p., 18 pls. Paris.

Lisson, C. I. (1907) *Contribución a la Geología de Lima y sus alrededores*, Lima.

—— (1930) *Trigonias neócomicos del Perú*, Bol. Minas, 2, 20, 26 p., 10 pls. Peru.

Lycett, John (1872–1879) *A monograph of the British fossils Trigoniae*, Palaeontographical Soc., 245 p., 41 pls.; Suppl., 1881–1883, 18 p., 4 pls. London.

Meek, F. B. and Hayden, F. V. (1865) *Paleontology of the upper Missouri; invertebrates*, Smithsonian Contributions to Knowledge, vol. 14, no. 172, 135 p.

Moore, Hilary B. (1932) *The faecal pellets of Anomura*, Royal Soc. Edinburgh, Session 1931–1932, Pro., vol. 52, pt. 3, no. 14, p. 296. Edinburgh.

Muller, Siemon Wm. and Schenck, Hubert G. (1943) *Standard of Cretaceous system*, Am. Assoc. Petrol. Geol., Bull., vol. 27, no. 3, p. 262–278. Tulsa.

Nagao, Takumi (1932) *Some Cretaceous mollusca from Japanese Saghalin and Hokkaido (Lamellibranchiata and Gastropoda)*, Hokkaido Imp. Univ., Jour. Faculty Sci., ser. 4, Geol. and Mineral., vol. 2, no. 1, p. 23–50, 4 pls. Sapporo.

Neumayr, M. and Uhlig, V. (1881) *Ueber Ammonitiden aus den Hilsbildungen Norddeutschlands*, Palaeontographica, Bd. 27, p. 7–75, i–iv, pls. 15–67. Cassel.

Nikchitch, J. (1915) *Représentants du genre Douvilleiceras de l'Aptien du versant septentrional du Caucase*, Com. Géol., Mém., Nouvelle Série, Livr. 121, 53 p., 6 pls. Petrograd.

Nikitin, S. (1888) *Les vestiges de la période crétacé dans la Russie centrale*, Com. Géol., Mém., vol. 5, no. 2, 205 p., 5 pls. St. Pétersbourg.

Orbigny, Alcide, d' (1840) *Paléontologie française. Terrain crétacé, vol. 1.—Céphalopodes*, 652 p., 148 pls. Paris.

―――― (1842) *Voyage dans l'Amérique méridionale, 3, pt. 4, Paléontologie*, 188 p., 22 pls. Paris and Strasbourg.

―――― (1843–1847) *Paléontologie française. Terrain crétacé, vol. 3.—Lamellibranches*, 807 p., 489 pls. Paris.

Pictet, F. J. and Campiche, G. (1858–1872) *Description des fossiles du terrain crétacé des environs de Sainte-Croix*, Matériaux pour la Paléontologie Suisse, ser. 2–6, vols. 1–5, p. 280–752–558–352–158, pls. 2–43–208. Genève and Bâle.

―――― **and Renevier, Eugène** (1858) *Descriptions des fossiles du terrain aptien de la Perte du Rhône et des environs de Sainte-Croix*, Matériaux pour la Paléontologie Suisse, ser. 1, 184 p., 23 pls. Genève.

Quenstedt, Werner (1930) *Die Anpassung an die grabende Lebensweise in der Geschichte der Solenomyiden und Nuculaceen*, Geol. u. Paläont. Abh., Bd. 22 (Neue Folge, Bd. 18), Heft 1, 119 p., 3 pls. Jena.

Ransome, Frederick Leslie (1904) *The geology and ore deposits of the Bisbee quadrangle Arizona*. U. S. Geol. Survey, Prof. Paper 21, 168 p., 29 pls. Washington.

Renngarten, V. (1926) *La faune dés dépôts crétacés de la region d'Assa-Kambiléevka, Caucase du Nord*. Com. Géol., Mém., Nouvelle Série, Livr. 147, 132 p., 9 pls. Leningrad.

―――― (1931) *The High Ingushetia. Geological explorations in the valleys of two rivers, Assa and Kambileevka, North Caucasus*. Geol. and Prospecting Service of U. S. S. R., Tr., fascicle 63, 192 p., 6 pls. Leningrad.

Riedel, L. (1937) *Amonitas del cretácico inferior de la Cordillera Oriental*. Estudios geól. y paleónt. sobre la Cordillera Oriental de Colombia, pt. 2, Depart. Minas y Pétrol., Ministerio de Industrias y Trabajo, Repúb. Colombia, p. 7–80, pls. 3–14.

Roemer, Ferdinand (1852) *Die Kreidebildungen von Texas und ihre organischen Einschlüsse*, 100 p., 11 pls. Bonn.

Roman, Frédéric (1938) *Les ammonites jurassiques and crétacées. Essai de genera*, 554 p., 53 pls. Paris.

Sarasin, Ch. (1897) *Quelques considérations sur les genres* Hoplites, Sonneratia, Desmoceras, *et* Puzosia, Soc. Géol. France, Bull., sér. 3, vol. 25, p. 760–799. Paris.

Schenck, Hubert G. (1934) *Classification of nuculid pelecypods*, Mus. Royale Hist. Nat. Belg., Bull., vol. 10, no. 20, 78 p., 5 pls. Bruxelles.

―――― (1936) *Nuculid bivalves of the genus* Acila, Geol. Soc. Am., Spec. Papers, no. 4, 149 p., 18 pls.

―――― (1939) *Revised nomenclature for some nuculid pelecypods*, Jour. Paleont., vol. 13, no. 1, p. 21–41, pls. 5–8.

―――― (1943) Acila princeps, *a new Upper Cretaceous pelecypod from California*, Jour. Paleont., vol. 17, no. 1, p. 60–68, pls. 8–9.

Schrader, Frank C. (1915) *Mineral deposits of the Santa Rita and Patagonia Mountains, Arizona*, U. S. Geol. Survey, Bull. 582, 373 p. Washington.

Schuchert, Charles (1935) *Historical geology of the Antillean-Caribbean region, or the lands bordering the Gulf of Mexico and the Caribbean Sea*, 811 p., John Wiley and Sons. New York.

REFERENCES

Scott, Gayle (1926) *Études stratigraphiques et paléontologiques sur les terrains crétacés du Texas*, Thèse présentée a la Faculté Sci. Grenoble, Univ. Grenoble, 218 p., 3 pls.

———— (1940) *Cephalopods from the Cretaceous Trinity group of the south-central United States*, Univ. Texas, Pub. 3945, p. 969–1106, pls. 56–68.

Seunes, J. (1887) *Notes sur quelques ammonites du Gault*, Soc. Géol. France, Bull., sér. 3, vol. 15, p. 557–571, pls. 11–14. Paris.

Sharpe, Daniel (1850) *On the secondary district of Portugal which lies on the north of the Tago*, Geol. Soc. London, Quart. Jour., vol. 6, p. 135–200, pls. 14–26.

Sinzow, J. (1906) *Die Beschreibung einiger Douvilléiceras-Arten aus dem oberen Neocom Russlands*, Russ.-K. Gesell. St. Petersburg, Verhandl., ser. 2, vol. 44, p. 157–197, pls. 1–5.

———— (1908 (1907)) *Untersuchung einiger Ammonitiden aus dem unteren Gault Mangyschlaks und des Kaukasus*, Russ.-K. Gesell. St. Petersburg, Verhandl., ser. 2, vol. 45, p. 455–519, pls. 1–8.

Smith, James Perrin (1932) *Lower Triassic ammonoids of North America*, U. S. Geol. Survey, Prof. Paper 167, 199 p., 81 pls. Washington.

Spath, L. F. (1921) *On Cretaceous cephalopoda from Zululand*, South African Mus., Ann., vol. 12, no. 16, p. 217–321, pls. 19–26.

———— (1922) *On Cretaceous ammonoids from Angola, collected by Professor J. W. Gregory*, Royal Soc. Edinburgh, Tr., vol. 53, pt. 1, no. 6, p. 91–160, pls. 1–4.

———— (1923a; 1925; 1931) *A monograph of the ammonoidea of the Gault*, Palaeontographical Soc., pt. 1, p. 1–72, pls. 1–4; pt. 2, p. 73–110, pls. 5–8; pt. 8, p. 313–378, pls. 31–36. London.

———— (1923b) *On the ammonite horizons of the Gault and contiguous deposits*, Geol. Survey Great Britain and Mus. Practical Geology, Summ. Prog. 1922, p. 139–149.

———— (1930) *On some ammonoidea from the Lower Greensand*, Ann. Mag. Nat. Hist., ser. 10, vol. 5, no. 29, p. 417–464, pls. 14–17. London.

———— (1931) *Revision of the Jurassic cephalopod fauna of Kachh (Cutch), pt. 5*, India Geol. Survey, Palaeont. Indica, n.s., vol. 9, Mem. no. 2, p. 551–658, pls. 103–124. Calcutta.

———— (1934) *The ammonoidea of the Trias*, Catalogue of the fossil cephalopoda, British Mus. (Nat. Hist.), pt. 4, 521 p., 18 pls. London.

———— (1935) *On the age of certain species of* Trigonia *from the Jurassic rocks of Kachh (Cutch)*, Geol. Mag., vol. 72, p. 184–189. London.

———— (1939a) *Problems of ammonite nomenclature. 5. On* Acanthohoplites jacobi *(Collet) and the jacobi zone of the Folkestone Sands*, Geol. Mag., vol. 76, p. 236–239. London.

———— (1939b) *Mesozoic ammonoidea*, Fortsch. Paläont., Bd. 2, p. 203–210. Berlin.

Stanton, T. W. (1897) *A comparative study of the Lower Cretaceous formations and faunas of the United States*, Jour. Geol., vol. 5, p. 579–624.

———— (1901) *The marine Cretaceous invertebrates*, Princeton Univ. expeditions to Patagonia, 1896–1899, Repts., vol. 4—Paleontology, pt. 1, 43 p., 10 pls.

Steinmann, Gustav (1881) *Zur Kenntniss der Jura- und Kreideformation von Caracoles (Bolivia)*, Neues Jahrb. Mineral., Geol. u. Palaeont., Beil-Bd. 1, p. 239–301, pls. 9–14. Stuttgart.

———— (1882) *Die Gruppe der Trigoniae pseudo-quadratae*, Neues Jahrb. Mineral., Geol., Palaeont., Beil-Bd. 1, p. 219–228, pls. 7–9. Stuttgart.

Stephenson, Lloyd William (1941) *The larger invertebrate fossils of the Navarro group of Texas*, Univ. Texas, Pub. 4101, 641 p., 95 pls. Austin.

Stolley, E. (1908a) *Die Gliederung der norddeutschen unteren Kreide*, Centralbl. Mineral., Geol., Paläont., Jahrg. 1908, p. 107–124; 140–151; 162–175; 211–220; 242–250. Stuttgart.

———— (1908b) *Zur Kenntniss der kaukasischen Unterkreide*, Centralbl. Mineral., Geol. Paläont., Jahrg. 1908, p. 321–325. Stuttgart.

Stoyanow, Alexander (1910) *On a new genus of brachiopoda*, Acad. Imp. Sci. St.-Pétersbourg, Bull., p. 853–855.

———— (1915) *On some Permian brachiopoda of Armenia*, Com. Géol., Mém., N. S., Livr. 111, 95 p., 6 pls. Petrograd.

———— (1936) *Occurrence of the Malone and Torcer faunas at the base of the Arizona Comanchean*, Science, n.s., vol. 83, p. 328.

Stoyanow, Alexander (1937) *Fossiliferous zones in the Cretaceous and Tertiary deposits of southeastern Arizona* (Abstract), Geol. Soc. Am., Pr., 1936, p. 296.

——— (1938) *Lower Cretaceous stratigraphy in southeastern Arizona* (Abstract), Geol. Soc. Am., Pr., 1937, p. 117.

——— (1942) *Paleozoic paleogeography of Arizona*, Geol. Soc. Am., Bull., vol. 53, p. 1255–1282, pls. 1–5. New York.

Taliaferro, N. L. (1933) *An occurrence of Upper Cretaceous sediments in northern Sonora, Mexico*, Jour. Geol., vol. 41, p. 12–37. Chicago.

Tenney, J. B. (1932) *The Bisbee mining district, In: Ore deposits of the Southwest*, 16th Intern. Geol. Cong., Guidebook 14, p. 40–67. Washington.

Uhlig, V. (1901) *Über die cephalopodenfauna der Teschener und Grodischter Schichten*, K. Akad. Wiss. Math.-Naturw. Kl., Denkschr. Bd. 72, 87 p., 9 pls. Wien.

——— (1910) *Die Fauna der Spiti-Schiefer des Himalaya, ihr geologisches Alter und ihre Weltstellung*, K. Akad. Wiss., Math.-Naturw., Kl., Denkschr. Bd. 85, p. 597–609. Wien.

Weaver, Charles E. (1931) *Paleontology of the Jurassic and Cretaceous of West Central Argentina*, Univ. Wash., Mem., vol. 1, 594 p., 62 pls. Seattle.

Weller, Stuart (1907) *A report on the Cretaceous paleontology of New Jersey*, Geol. Survey N. J., Paleont. Ser., vol. 4, 871 p. (text), 111 pls. (atlas). Trenton.

Whitfield, Robert P. (1885) *Brachiopoda and lamellibranchiata of the Raritan clays and Greensand marls of New Jersey*, U. S. Geol. Survey, Mon. 9, 338 p., 35 pls. Washington.

Woodring, W. P. (1927) *American Tertiary mollusks of the genus* Clementia, U. S. Geol. Survey, Prof. Paper 147-C, p. 25–47, pls. 14–17. Washington.

Woods, Henry (1899–1903; 1904–1913) *A monograph of the Cretaceous lamellibranchia of England*, Palaeontographical Soc., vol. 1, 232 p., 42 pls.; vol. 2, 473 p., 62 pls. London.

EXPLANATION OF PLATES 8–26

PLATE 8.—*ACILA, IDONEARCA*

Figure | Page
1–8. *Acila (Truncacila) schencki*, sp. nov. Pacheta member, D. 9g...................... 61
  (1). Syntype. No. 91029. Left valve. ×1.1.
(2–3). Syntype. No. 91030. Right valve. ×1.1; and enlargement to show details of ornamentation. ×2.4.
  (4). Syntype. No. 91053. Interior of right valve in a slab of Lancha limestone. ×1.7.
  (5). Enlargement of preceding to show dentition, adductor impressions, and position of embedded faecal pellets. ×2.4.
  (6). Detail of preceding showing the pit on the spoon-shaped platform and the trigonal resilium pit. ×7.
  (7). Faecal pellets illuminated to show the paired lobes. ×2.6.
  (8). Enlargement of preceding. The right pellet is inclined and partly shows the lateral sides of lobes. A part of the third pellet is seen above. ×8.5.
9–12. *Idonearca stephensoni*, sp. nov. Pacheta member, D.9g............................ 63
  (9). Syntype. No. 92156. Right valve showing anterior ribs. ×1.1.
(10). Syntype. No. 92141. Left valve showing dentition. ×1.
(11). Syntype. No. 91038. Left valve with ribbing in the posterior part. ×1.1.
(12). Syntype. No. 92140. Left valve showing dentition and ligamental area. ×1.1.

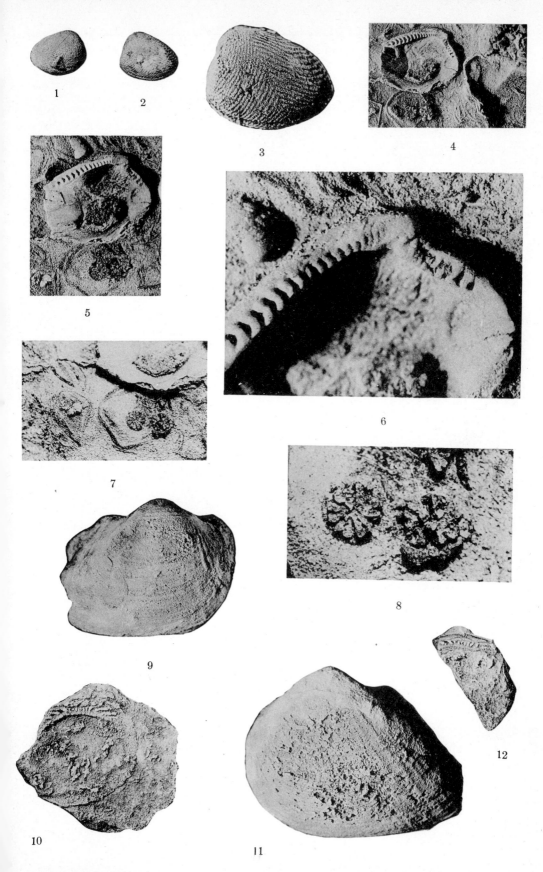

ACILA, IDONEARCA

IDONEARCA, LIMA, PTERIA, GERVILLIA

PLATE 9.—*IDONEARCA, LIMA, PTERIA, GERVILLIA*

Figure / Page

1. *Idonearca stephensoni*, sp. nov. Pacheta member, D.9g. Syntype. No. 91831. A ribless right valve. ×1 63
2. *Lima muralensis*, sp. nov. Base of Mural limestone. Holotype. No. 94010. Right valve. ×1.4 93
3-4. *Lima cholla*, sp. nov. Cholla member, D.5e 93
  (3). Paratype. No. 91467. Left valve showing the ribless anteroventral region. ×1.
  (4). Holotype. No. 91491. Left valve. ×1.2.
5. *Pteria peregrina*, sp. nov. Perilla member, D.2b. Holotype. No. 91841. Left valve. ×1.1 65
6-7. *Gervillia heinemani*, sp. nov. Saavedra member, D.7i. Holotype. No. 91100. Left valve. ×1 64
8. *Lima espinal*, sp. nov. Joserita member, D.8c. Holotype. No. 91181. Left valve. ×1 92

PLATE 10.—*GERVILLIA, EXOGYRA*

Figure / Page

1. *Gervillia cholla*, sp. nov. Cholla member, D.6d. Holotype. No. 91879. Left valve. ×1 64
2-3. *Gervillia rasori*, sp. nov. Cholla member, D.6d 65
(2). Paratype. No. 91958. Left valve. ×0.8.
(3). Holotype. No. 91548. Right valve. ×1.
4-6. *Exogyra lancha*, sp. nov. Pacheta member, D.9g 66
(4). Holotype. No. 91054. Left valve. ×1.
(5). Paratype. No. 91055. Left valve. ×1.3.
(6). Paratype. No. 91056. Left valve. ×1.1.
7-8. *Exogyra*, sp. Pacheta member, D.9g. Specimens. Nos. 91061 and 91063. Left valves. ×1 66

GERVILLIA, EXOGYRA

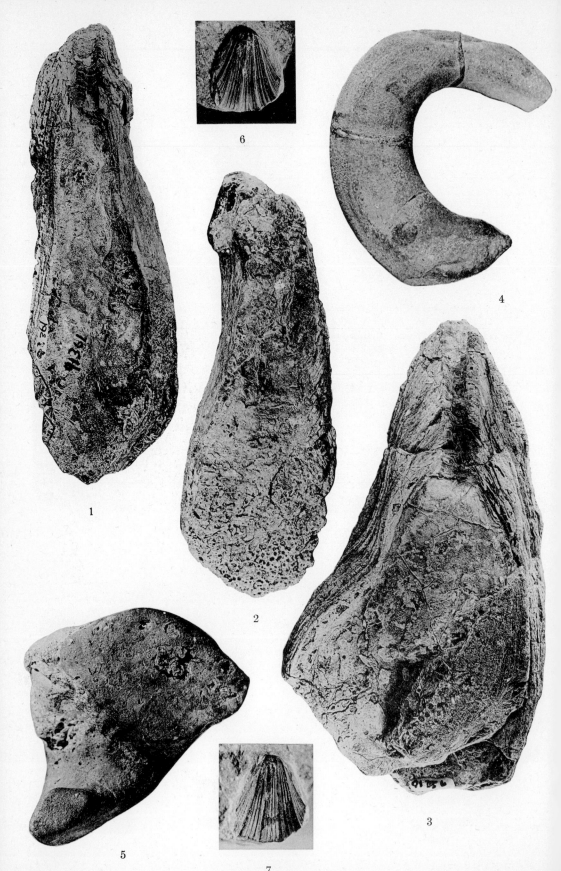

OSTREA, CAPRINA, NEITHEA

PLATE 11.—*OSTREA, CAPRINA, NEITHEA*

Figure	Page
1–3. *Ostrea edwilsoni*, sp. nov. Saavedra member, D.7d.	66
(1). Syntype. No. 91361. Interior of the left valve. A narrow variety. ×1.	
(2). Same as preceding. Exterior of the left valve. ×1.	
(3). Syntype. No. 91356. Interior of the left valve. A wide variety. ×1.	
4. *Caprina*, sp. Specimen. No. 94001. Basal member of the Mural limestone. Little hills north of Hay Flat. ×1.	94
5. *Caprina*, sp. Specimen. No. 94002. Same horizon and locality. ×1.2.	94
6–7. *Neithea vicinalis*, sp. nov. Perilla member, D.2b. Holotype. No. 91479. Two differently illuminated views of the right valve. ×1.5.	91

PLATE 12.—*TRIGONIA*

Figure   Page

1-2. *Trigonia guildi*, sp. nov. Cholla member, D.6a 75
  (1). Holotype. No. 91205. Left valve. Paired costellae in the posterior of the areas. $\times 1.1$.
  (2). Same as preceding. Escutcheonal tubercles are arranged in transverse rows in the advanced stage. $\times 1.1$.
  3. *Trigonia resoluta*, sp. nov. Cholla member, D.6a. Holotype. No. 91418. Left valve. Resolved costellae in the anterior part of the outer area. $\times 1$ 76
  4. *Trigonia dumblei*, nom nov. Malone, Texas. Originally described by Cragin as "*Trigonia taffi*" and later transferred to "*Trigonia vyschetzkii*." Holotype. U. S. National Museum. Left valve showing resolved costellae in the anterior part of the outer area. $\times 1$ 77
5-6. *Trigonia mearnsi*, sp. nov. Perilla member, D.2b. Guadalupe Canyon 78
  (5). Paratype. No. GU47. Left valve of a young form showing W-shaped umbonal ribs and resolved costellae on both areas. $\times 1$.
  (6). Dorsal view of preceding showing V-shaped initial umbonal ribs. $\times 1$.

TRIGONIA

TRIGONIA

EXPLANATION OF PLATES

PLATE 13.—*TRIGONIA*

Figure	Page

1–4. *Trigonia mearnsi*, sp. nov. Perilla member, D.2b. Guadalupe Canyon.............. 78
  (1). Holotype. No. GU46. Right valve. Both areas with resolved costellae except in the posterior third. ×1.
  (2). Paratype. No. GU50. Dorsal view showing escutcheonal tubercles set obliquely along the lines of growth. ×1.
  (3). Paratype. No. GU45. Dorsal view showing umbonal ribs and resolved areal costellae. ×1.
  (4). Paratype. No. GU42. A somewhat weathered younger form. Right valve. ×1.
5. *Trigonia saavedra*, sp. nov. Saavedra member, D.7n. Holotype. No. 91333. Left valve showing areal costellae resolved anteriorly and paired posteriorly. ×1.1...... 78
6–10. *Trigonia cragini*, sp. nov. Cholla member, D.6a, d, f............................ 80
  (6). Syntype. No. 91509. Cholla member, D.6a. Left valve. Showing umbonal costation, initial marginal carina, and undifferentiated area. ×1.8.
  (7). Syntype. No. 91429. Cholla member, D.6a. Left valve. ×1.1.
  (8). Syntype. No. 91445. Cholla member, D.6d. Right valve. ×1.1.
  (9). Syntype. No. 91434. Cholla member, D.6f. Left valve. ×1.2.
  (10). Syntype. No. 91430. Cholla member, D.6a. Shows gradual development of the outer area, median furrow, and escutcheonal ornamentation. Note individualization of the inner area between the escutcheon and the median furrow in the posterior part of the specimen. ×2.

PLATE 14.—*TRIGONIA*

Figure — Page

1. *Trigonia calderoni* (Castillo and Aguilera?) Cragin. A homoeotype collected by Stanton from the base of Truncate Mound, Malone, Texas. U. S. National Museum. Plaster cast of the left valve. ×1.1... 82

2. Same as preceding. Dorsal view. Shows normal V-s on the flank, very small inverted V-s on the initial marginal carina, striated outer area, and development of the median furrow and the inner area, in the posterior part. ×1.2........................ 82

3. *Trigonia cragini*, sp. nov. Same specimen as illustrated in Figure 8 of Plate 13. Shows encroachment of inverted V-s upon the marginal carina and differentiation of areas in the posterior part. ×1.9.. 80

4–10. *Trigonia kitchini*, sp. nov. Cholla member, D.6a............................... 82
  (4). Syntype. No. 91449. Right valve. ×1.
(5–6). Syntype. No. 91603. Right valve and escutcheonal view. ×1.
(7–8). Syntypes. Nos. 91505 and 91506. Details of escutcheonal ornamentation. ×1.
(9–10). Syntype. No. 91507. Left valve. Shows initial umbonal ribs and gradual development of the marginal carina, outer area, median furrow, and inner area. ×1.6.

11–14. *Trigonia reesidei*, sp. nov. Pacheta member, D.9g. Holotype. No. 91049. Right valve, left valve, umbonal view, and escutcheonal view. ×1..................... 83

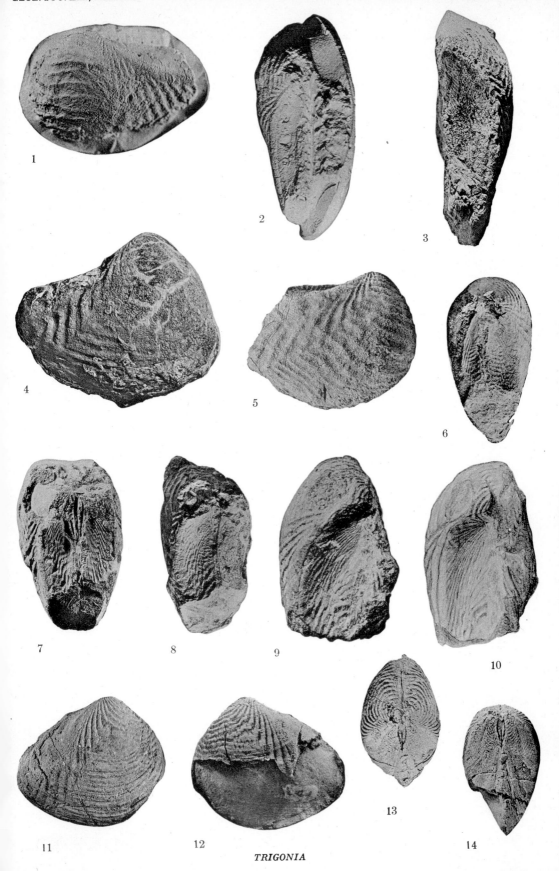

TRIGONIA

TRIGONIA, ASTARTE

PLATE 15.—*TRIGONIA, ASTARTE*

Figure — Page

1–3. *Trigonia reesidei*, sp. nov. Pacheta member, D.9g.............................. 83
  (1). Enlarged escutcheonal view of the holotype. Showing marginal carina, outer area, median furrow, inner area, escutcheon, and broadly V-shaped riblets. ×1.8.
  (2). Syntype. No. 91050. Left valve. A young form showing acute initial marginal carina. ×1.
  (3). Syntype. No. 91051. Right valve. Shows details of early costation. ×1.1.
4–8. *Trigonia weaveri*, sp. nov. Quajote member, D.3d............................... 87
  (4). Syntype. No. 91997. Left valve. ×1.
  (5). Syntype. No. 91245. Left valve (inverted print). ×1.
  (6). Syntype. No. 91200. Right valve. Shows ribs undulating anteriorly and dwindling posteriorly. ×1.
  (7). Same as Figure 5. To show umbonal ridge which outlines a triangular space in front. ×1.1.
  (8). Same as preceding. To show umbonal ridge, marginal angulation, median furrow which separates the areas, normal V-s on the inner area, and smooth escutcheon. ×1.1.
9–11. *Trigonia stolleyi* Hill. Perilla member, D.2b. Guadalupe Canyon................. 88
(9–10). Plesiotype. No. GU33. Right valve and escutcheonal view. ×1.
  (11). Plesiotype. No. GU20. Escutcheonal view. Showing differentiation of the areas. ×1.
12–14. *Astarte adkinsi*, sp. nov. Perilla member, D.2b. Guadalupe Canyon............... 93
  (12). Syntype. No. GU27. Right valve. ×1.2.
  (13). Syntype. No. GU32. Left valve. ×1.
  (14). Syntype. No. GU19. Interior of the right valve. ×1.2.

PLATE 16.—*TRIGONIA, UNICARDIUM, PECTEN*

Figure / Page

1–3. *Trigonia* sp. ex *aliformis* group. Perilla member, D.2b............................ 89
(1–2). Specimen. No. 91740. Ninety One Hills area. Left valve and dorsal view. ×1.1.
(3). Specimen. No. GU72. Guadalupe Canyon. Left valve. Shows ribbing and posterior separation of the areas by a wide median furrow. ×1.

4. *Unicardium*, sp. Specimen. No. 91811. Quajote member, D.3b. Right valve. ×0.9... 95

5–8. *Pecten (Chlamys) thompsoni*, sp. nov. Saavedra member, D.7d.................... 90
(5). Syntype. No. 91400. Right valve of a young form with preserved ears. ×1.2.
(6–7). Syntype. No. 91955. Left and right valves showing details of ornamentation on the flanks. ×1.1.
(8). Syntype. No. 91339. Right valve showing ornamentation of the ears. ×1.1.

TRIGONIA, UNICARDIUM, PECTEN

PECTEN, KAZANSKYELLA, PARAHOPLITES

PLATE 17.—*PECTEN, KAZANSKYELLA, PARAHOPLITES*

Figure	Page
1. *Pecten (Chlamys) thompsoni*, sp. nov. Syntype No. 91951. Left valve. Saavedra member, D.7d. ×1	90
2–8. *Kazanskyella arizonica*, sp. nov. Pacheta member, D.9c	100

(2). Syntype. No. 91748. ×1.
(3–4). Syntype. No. 91749. ×1.2.
(5–6). Syntype. No. 91747. ×1.1.
(7–8). Syntype. No. 91111. ×1.1; and enlargement to show details of the suture. ×2.6.
9–10. Suture of *Parahoplites "melchioris"* Sinzow (1908, Pl. 2, figs. 2 and 4) from Mangyshlak. From original photographs by Sinzow.

PLATE 18.—*SINZOWIELLA, BEUDANTICERAS*

Figure	Page
1–17. *Sinzowiella spathi*, sp. nov. Pacheta member, D.9c.	102

(1–5). Syntypes. Nos. 91203, 91083, 91099, 91074, 91112. Showing development of the costation. ×1.
(6–8). Syntype. No. 91113. To show ornamentation on the flank and venter, and whorl section at an advanced stage. ×1.
(9). Syntype. No. 91087. To show costation at an advanced stage. ×1.
(10). Syntype. No. 91210. Showing first lateral lobe at an advanced stage. ×1.
(11–12). Syntype. No. 91110. ×2; and enlargement to show external lobe at an early stage. ×4.8.
(13–14). Same as preceding. ×2; and enlargement to show parts of the suture at an early stage. ×2.9.
(15–16). Syntype. No. 91095. ×2.2; and enlargement showing trifid first lateral lobe and bifid saddles at an early stage. ×3.
(17). Syntype. No. 91100. To show suture at an advanced stage. ×1.

18–21. *Beudanticeras victoris*, sp. nov. Quajote member, D.3d.	127

(18). Holotype. No. 91909. ×1.1.
(19–20). Same as preceding. Part of the last whorl removed to show whorl sections and venter. ×1.1.
(21). Same as preceding. Showing falciform striae on the flank and sigmoidal bulges on the inner whorl. ×1.1.

SINZOWIELLA, BEUDANTICERAS

ACANTHOHOPLITES

PLATE 19.—*ACANTHOHOPLITES*

Figure   Page
1–13. *Acanthohoplites schucherti*, sp. nov. Quajote member, D.3d...... 109
(1–2). Holotype. No. 91199. ×1.1.
(3). Part of the holotype showing details of ribbing on the last whorl. ×1.5.
(4–5). Part of the holotype showing whorl sections. ×1; and enlargement. ×1.5. Lower margin of the specimen is damaged.
(6). Part of the holotype showing strong crest-like lateral tubercles. ×1.2.
(7). Inner whorls of the holotype showing straight ribs and migration of lateral tubercles. ×1.
(8). Part of the holotype showing first lateral lobe. ×1.7.
(9–10). Early whorl of the holotype. ×1.9; and enlargement to show lateral tubercles at the peripheral margin, bifurcating ribs, elongate ventral bullae, and bifid external saddle. ×4.5.
(11–13). Paratype. No. 91675. Detached whorl of the paratype to show details of the suture, whorl sections, and flattening of the venter. ×1.
14–16. *Acanthohoplites berkeyi*, sp. nov. Quajote member, D.3b...... 111
(14–15). Holotype. No. 92230. ×1.
(16). Paratype. No. 92249. ×1. Densicostate inner whorls.
17–20. *Acanthohoplites impetrabilis*, sp. nov. Quajote member, D.3b...... 112
(17–18). Syntype. No. 92228. ×1.
(19–20). Syntype. No. 92198. ×1.
21–23. *Acanthohoplites erraticus*, sp. nov. Quajote member, D.3b...... 113
(21–22). Holotype. No. 92500. ×1.
(23). Detail of preceding showing first lateral lobe. ×1.6.

PLATE 20.—*ACANTHOHOPLITES, IMMUNITOCERAS*

Figure	Page
1–6. *Acanthohoplites hesper*, sp. nov. Quajote member, D.3b........................	115

(1–2). Syntype. No. 92226. ×1.1.
(3–6). Syntype. No. 92227. ×1. Figure 3 shows gradual migration of lateral tubercles and substitution with umbilical bullae. Figure 6 shows first lateral lobe.

    7. *Acanthohoplites teres*, sp. nov. Quajote member, D.3b. Holotype. No. 92229. ×1... 114

8–15. *Immunitoceras immunitum*, sp. nov. Quajote member, D.3b.................... 117

(8–9). Holotype. No. 92563. Penultimate whorl. ×1.
(10). Same specimen showing whorl section. ×1.2.
(11–12). Parts of the holotype showing whorl section and venter. ×1.
(13). Complete holotype. ×1.1.
(14–15). Paratype. No. 92231. ×1.

   16. *Acanthohoplites* sp. No. 92571. Large specimen showing first lateral lobe. ×1.... 106

ACANTHOHOPLITES, IMMUNITOCERAS

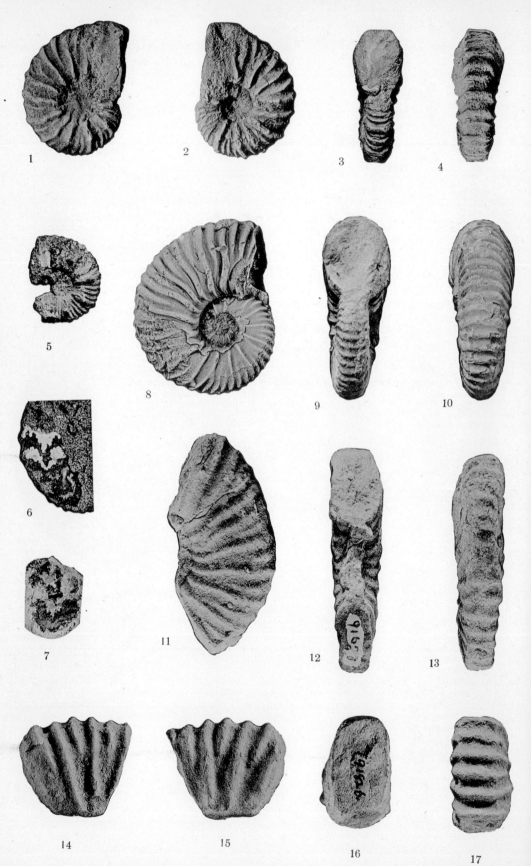

PARACANTHOPLITES, COLOMBICERAS ?, DUFRENOYA

PLATE 21.—*PARACANTHOHOPLITES, COLOMBICERAS ?, DUFRENOYA*

Figure   Page

1–7. *Paracanthohoplites meridionalis*, sp. nov.   Joserita member, D.8e.................. 118

(1). Holotype. 91150. Right flank. ×1.1.

(2–4). Same as preceding. Left flank, whorl section, and ventral view. ×1.

(5). Paratype. No. 91170. ×1.

(6). Paratype. No. 91153. ×1.5. To show partly preserved suture.

(7). Paratype. No. 91164. ×1.7. To show partly preserved first lateral lobe.

8–10. *Colombiceras? brumale*, sp. nov. Quajote member, D.3d. Holotype. No. 91693. ×1.. 122

11–17. *Dufrenoya justinae* (Hill). Cholla member, D.5g.............................. 124

(11–13). Plesiotype. No. 91653. ×1.

(14–17). Plesiotype. No. 91526. ×1.

PLATE 22.—*DUFRENOYA?*

Figure	Page
1–8. *Dufrenoya? compitalis*, sp. nov. Cholla member, D.5g	125
(1–2). Syntype. No. 91645. ×1.	
(3). Syntype. No. 91661. ×1.	
(4–5). Syntype. No. 91547. ×1.	
(6–8). Syntype. No. 91646. ×1.	

DUFRENOYA?

DUFRENOYA?, DUFRENOYA, SINZOWIELLA?

EXPLANATION OF PLATES

PLATE 23.—*DUFRENOYA?, DUFRENOYA, SINZOWIELLA?*

	Page
Figure	
1–3. *Dufrenoya? compitalis*, sp. nov. Cholla member, D.5g	125
(1–2). Syntype. No. 91683. ×1.	
(3). Syntype. No. 91662. ×1.	
4–6. *Dufrenoya joserita*, sp. nov. Joserita member, D.8d	125
(4–5). Syntype. No. 91142. ×1.	
(6). Syntype. No. 91169. ×1.	
7–13. *Sinzowiella?* sp. Pacheta member, D.9d	103
(7–9). Specimen. No. 91202. Fragment. Showing flank, whorl section, and venter. ×1.	
(10–11). Specimen. No. 91126. Lateral and ventral views. ×1.	
(12–13). Specimens. Nos. 91209 and 91092. Two conspecific fragments. ×1.	

PLATE 24.—*DESHAYESITES?, CLEONICERAS?*

Figure   Page
1–3. *Deshayesites? temerarius*, sp. nov. Specimen. No. 91222. 1½ miles northeast of Black Knob Hill. Probably above the Espinal grit. Left flank, right flank, and whorl sections. ×0.7.. 126
4–5. *Cleoniceras? schlaudti*, sp. nov. Holotype. No. 91068. Dry wash southeast of Schlaudt Ridge. Lateral and front views. ×0.6... 130

DESHAYESITES?, CLEONICERAS?

DESHAYESITES?, ACANTHOHOPLITES, COLOMBICERAS?

EXPLANATION OF PLATES

PLATE 25.—*DESHAYESITES?, ACANTHOHOPLITES, COLOMBICERAS ?*

Figure / Page

1-3. *Deshayesites? butleri*, sp. nov. Joserita member, D.8c. Holotype. No. 91204. Whorl section, flank, and venter. ×0.8.. 126

4. Suture of *Acanthohoplites trautscholdi* (Simonovitsch, Sorokin, and Bazewitsch). Homoeotype from Mangyschlak. From original photograph by Sinzow (1908, Pl. 4, fig. 15).. 106

5. Specimen. No. 91694. Fragment of a flat-ribbed ammonite collected with *Colombiceras? brumale* from Cholla member, D.3d. ×1.1............................... 122

PLATE 26.—*DOUVILLEICERAS?, STOLICZKAIA*

Figure   Page
1–2. *Douvilleiceras? muralense*, sp. nov. Mural limestone. Mural Hill Ridge............. 130
(1). Specimen. No. 94029. Part of the penultimate whorl showing external lobe and external saddle. ×0.6.
(2). Same as preceding. Whorl section. ×0.8.
3–4. *Stoliczkaia patagonica*, sp. nov. Molly Gibson formation. Patagonia Mountains. Holotype. No. PT23. Flank and venter. ×0.9............................... 128
5–6. *Stoliczkaia excentrumbilicata*, sp. nov. Molly Gibson formation. Patagonia Mountains. Holotype. No. PT01. Flank and venter. ×1............................... 129
7–8. *Stoliczkaia scotti*, sp. nov. Molly Gibson formation. Patagonia Mountains. Holotype. No. PT10. Flank and venter. ×0.9............................... 129

DOUVILLEICERAS?, STOLICZKAIA

INDEX*

Abich, H., 35
abichi, subzone, 34
Abrupta group, 40, 56, 58
Acanthoceras, 31, 96
   milletianum var. *plesiotypica*, 119
"Acanthoceras," 5
   milletianum, 120
   — var. *elegans*, 32
   — var. *plesiotypica*, 32
Acanthohoplitan fauna, in Quajote, 21
Acanthohoplites, 32, 33, 37, 51, 52, 54, 95, 98, 103, 105, 106, 107, 108, 109, 113, 119, 120, 121, 149, 150, 155
   abichi, 38, 107, 108, 111
   aegis Anderson, 54
   aschiltaensis, 33, 36, 38, 107, 108, 112
   — genotype, 106
   — subzone, 35
   bergeroni, 38
   berkeyi, 22, 36, 54, **111**, 112, 149; Pl. 19, figs. 14–16
   — group, 108
   bigoureti, 32, 106, 107, 111
   — *abichi*, group, 108
   campichei, 106, 115, 116
   — correct spelling, 96
   erraticus, 36, **113**, 114, 149; Pl. 19, figs. 21–23
   — distribution, 13
   evolutus, 107, 108, 114
   hesper, 22, 33, 36, 54, 106, 109, **115**, 116, 150; Pl. 20, figs. 1–6
   — distribution, 13
   — group, 109
   impetrabilis, 22, 54, **112**, 113, 149; Pl. 19, figs. 17–30
   — distribution, 13
   — group, 108
   immunitus, 109
   jacobi, subzone, 35
   — zone, 35
   laticostatus, 107, 121, 122
   meridionalis, group, 109
   multispinatus var. *tenuicostata*, 119
   nolani, subzone, 35
   pulcher, 116
   — in *jacobi* zone, 32
   schucherti, 16, 19, 22, 36, 54, **109**, 111, 149; Pl. 19, figs. 1–13.
   — distribution, 13
   — range, 18

Acanthohoplites (Cont'd.)
   — Sinzow, 106
   — sp., 150
   subangulatus, 113
   subpeltoceroides, 121
   ? *subpeltoceroides*, 37, 38
   — suture of syntypes, 136
   teres, 22, 36, 54, 107, 108, **114**, 150; Pl. 20, fig. 7
   — distribution, 13
   — group, 108
   tobleri, 107, 111, 112
   trautscholdi, 155
   — zone, 39, 56
"*Acanthohoplites*"
   jacobi, 32, 120
   plesiotypicus, 32, 120
   tobleri, 121
   — var. *discoidalis*, 111
Acanthohoplitinae, 22, 32, 33, 49, **51**, 56, 95, 116, 121
   — subf. nov., 17, 103
Acila, 1, 55, 56, 61, 62, 138
   bivirgata, 61, 62
   castrensis, 63
   conradi, 62
   demessa, 62
   — oldest species described, 14
   schencki, 10, 19, 56, 62, 63
   — oldest species, 1
Acila (*Truncacila*), 10, 16, 61
   bivirgata, 56
   — in Folkestone Gault, 14
   castrensis, 62, 63
   conradi, 62
   decisa, 62
   hokkaidoensis, 62
   princeps, 62
   schencki, 22, 55, **61**, 63, 138; Pl. 8, figs. 1–8
   — in Lancha, 14
   shumardi, 62
   — pellets of, 63
Actaeonella, 5
   — sp. cf. *dolium*, 5
Adkins, W. S., 3, 43, 131
   — on Glen Rose, 23
   — on Malone and Torcer, 6
   — on red sandstones, 23
   — on "*Trigonia steeruwitzi*" Cragin, 5
   — on Wichita Paleoplain, 40

---
* Figures in **boldface** indicate detailed description.

Adobe Canyon, 58
Africa, northwestern, 56
Africano-Indian province, 69
Agassiz, L., 131
Aguilera, J. G., 41, 131
Albian, 31, 36
— ammonites, interrupted ribbing, 32
— base, 32
— lower boundary, 1
— transgression, 5
Albritton, C. C., 131
— on Malone sequence, 42
— on "Torcer formation," 43
Alexis, C., 30
Altar Headland, 53
Ammonites, Caucasian-Transcaspian relationship, 50
— communication route, 50
— migration, 50
Ammonites
  alexandrinus, 121
  bicurvatus, 97
  campichei, 98, 109, 115
  cornuelianus, 96, 104
  crassicostatus, 112, 121
  dispar, 128
  dufrenoyi, 124
  mammillatus, 96, 104
  martini, 96, 104
  milletianus, 32, 33, 119, 120, 121
  nodosocostatus, 96, 104
  royerianus, 104
  treffryanus, 112, 121, 122
Ammonitoceras tovilense, subzone, 35
Ammonoidea, 95
Anderson, F. M., 55, 131
— on Cretaceous provinces, 54
Anthula, D. J., 32, 35, 54, 98, 116, 131
— on Aptian, 31
— on *Parahoplites*, 96
Antuco, volcano, Chile, 46
Aptian, 36
— Albian, ammonite zones, 20
— basin, Texas, 1
— boundary, 20, 31
  — on natural foundation, 2
  — on paleontological grounds, 1
— lagoonal deposits, 52
— seaways, 57
— sequence, 35
— strata, Venezuela, 14
— time, termination, 1, 33
— transgression, 5
— zonal sequence, 33

Aptain (Cont'd.)
— zonation, 34
Aptian-Lower Albian zones, 34
Aptian, Old World, 1
Araucarioxylon, 11, 15, 16
Argentina, 65
Arivechi, Sonora, 52
Arizona, ammonite succession correlated, 20
— Cretaceous sequence, 1
— index zones, correlation, 36
— paleogeography, 1
Arkansas, Hill's types from, 30
Arkell, W. J., 56
— cited, 56
Arkill limestone, 11, 15
— last recurrence of *Trigonia reesidei*, 15
aschiltaensis, subzone, 34
— zone, 37
Aspidoceras laevigatum, 43
Astarte, 15, 16, 17, 93, 145
  adkinsi, 9, 17, 22, 93, 145; Pl. 15, figs. 12–14
  — distribution, 13
  — in Guadalupe Canyon, 29
  elongata, 94
Astartidae, 93
Astrocoenia sp., 5
"*Astrocoenia*," 10, 16
Atlantic border, 52
Atlantic region, 52
Austin, Cragin's collection, 2

Backlund, H. G., 130
Baga limestone, *Paracanthohoplites* in, 15
— shale, 12
Baker Springs, 28
Barata limestone, 11, 21
— new fauna introduced, 15
Basin Range structure, 57
Bedoulian, 36
Bettmar, 31
Beudanticeras, 51, 52, 95, 127, 130, 148
  hatchetense, 21
  stoliczkai, 127
  victoris, 16, 19, 21, 22, 36, **127**, 148; Pl. 18, figs. 18–21
  — distribution, 13
  — in Quajote, 9
  — range, 18
  walleranti, 127
Bigot, A., 131
Bisbee, 6, 8, 15, 23, 27, 28, 29, 50
— anticline, 7
— area, 6, 18, 19, 23, 24, 26, 27, 28
— Arizona, 2

Bisbee (Con'd.)
— beds, 5
— collecting localities, 6
— group, 6, 8, 23, 29, 55, 58
  — volcanic activity, 31
— Junction, 8, 18
— quadrangle, 6, 7, 23, 26
Black Knob, 26
— correct location, 26
— dolomite, 12, 27
— quartzite, 12, 14, 27
  — large trees in, 14
  — plant remains, 14
Black Knob Ridge, 14, 27
— quartzites, 24
— section, 8
— summit, 26
Bluff Mesa, 77
bodei, zone, 34
Bogdanowitsch, C., 131
Böse, E., 51, 52, 131
bowerbanki, subzone, 34
Branco, W., 48
Brewery Gulch, 6
British India, 70
Brown, B., 59
Buccinopsis, 15
Buch, L. v., 83, 84, 85, 131
Buda, 31
Burckhardt, C., 43, 51, 52, 65, 81, 125, 131
— on *Acanthohoplites*, 107
— on Bisbee group, 5
— on Malone strata, 41
— on Morita age, 23
Burnet, 39
Butler, B. S., 3, 29, 131
Butler, G. M., 3

Campiche, G., 134
Canille, 30
Cañon de Vallas, 51
Caprina occidentalis, 94
— sp., 20, 28, **94**, 141; Pl. 11, figs. 4–5
"*Caprina*" sp. cf. *occidentalis*, 5
Caprinidae, 94
Caprotina, 4
Carr Peak, 30
Casa Blanca Canyon, 59
Cassiope, 15
Castillo, A. D., 41, 131
Catorce, Mexico, 41, 44
Caucasus, 1, 2, 54, 55
— index zones, 20
Caucasian-Transcaspian affinities, 52

Cavins, O. A., 51, 131
Chao, Y. T., 131
Chapman, T. G., 3
Chapparal sandstone, 11
— tree remains in, 14
Cheloniceras, 32, 37, 96, **104, 105, 106, 107, 108**
  cornuelianum, 37, 104
  — group, 105
  ? *gottschei*, in Zululand, 108
  hambrovi, subzone, 35
  martini, group, 105
  — zone, 95
  — var. *caucasicum*, 37, 38
  royerianum, 104
  seminodosum, 105
  subnodosocostatum, 105
  — zone, 35
  tschernyschewi, 105
Cheloniceratidae, 32, 33, 104, 112, 130
— in America, 32
Chihuahua, 50
China, eastern, 55
Cholla member, 10, 13, 16, 18, 21, 22, 58
— ammonites in, 14
— analyzed, 16
— described, 10
— Texas fauna in, 1
Cienda limestone, 12
— parahoplitan fauna in, 14
Cintura formation, 20, 29, 31, 53
— absence, 28
— correlated with Albian, 5
— resembling Morita, 29
Clansayan, 36
— age, 35
— placed in Aptian, 1
— transitional fauna, 32
Clansayes, 31
— ammonite faunas, 31
— beds, 31
Clavellatae, 46, 68
Clementia, 16
  (*Flaventia*) *ricordeana*, 28
— sp., 28
Cleoniceras
  baylei, 130
  cleon, 97
  ? *schlaudti*, **130**, 154; Pl. 23, figs. 4–5
Clyde, 63
Coahuila, 52, 58
— northeastern, 58
— southeastern, 50
Coahuila-Durango region, 50
Cochise, 27

Collet, L. W., 32, 33, 131
Colombiceras, 95, 106, 112, 121
— aff. *tobleri*, 112
  crassicostatum, 112, 121
— group, 121
— Spath, 121
  tobleri var. *discoidalis*, 112
  treffryanum, group, 121
Colombiceras?, 151, 155
  brumale, 16, 22, **122,** Pl. 21, figs. 8–10
  — distribution, 13
  — range, 18
  treffryanum, 122
Comanche sea, 29
— series, 5
Conrad, T. A., 2, 131, 132
consobrinoides, subzone, 34
Coquand, H., 83, 132
Corta sandstone, 11, 15
'Cosmoceratida,' 104
Cragin, F. W., 2, 5, 6, 42, 70, 73, 77, 81, 82, 132
— on age of Malone fauna, 41
Cuchillo, 57
— formation, 39
Cucullaea, 15, 16, 17
Cuprite Mine, 60
Cyprimeria, 16, 17
  sp., 5

Daghestan, 35, 55
— ammonite stratigraphy of, 36
Dana, J. D., 2, 85, 132
Dane, C. H., 132
Darrow, R. A., Mrs., 3
Darton, N. H., 27, 57, 132
Deshayesites, 95, 126
  bodei, subzone, 35
  consobrinoides, subzone, 35
  consobrinus, 124, 126
  dechyi, 37, 38
  deshayesi, 36, 38, 97, 123
  — time, 23
  — zone, 34, 35, 37
  grandis, 126
— Kazansky, 125
  weissi, 123, 126, 127
  — subzone, 35
Deshayesites?, 154, 155
  butleri, 22, **126,** 155; Pl. 25, figs. 1–3
  temerarius, **126,** 154; Pl. 24, figs. 1–3
Deshayesitinae, 95
— subfam. nov., 123
"*Desmoceras* (*Uhligella*)" *walleranti*, 127

Desmoceratidae, 32, 95, 127
Deussen, A., 132
"*Diadochoceras*," 32
  subnodosocostatum, 32
Dietrich, W. O., 76, 83, 132
Djulfa, 55
Dos Cabezas Mountains, 27
Douglas, 6, 28, 60
Douvillé, H., 132
Douvilleiceras, 37
  cornueli, 104
  inaequinodum, 130
  — subzone, 35
  mammillatum, 37
  — subzone, 35
  — zone, 23, 31, 35, 36, 38, 39
Douvilleiceras?, 156
  muralense, 36, **130,** 156; Pl. 26, figs. 1–2
"*Douvilleiceras*," 96, 106
  subnodosocostatum, 31, 32
Dragoon Mountains, 27
Dufrenoya, 1, 23, 37, 56, 95, 123, 153
— Burckhardt, 124
  dufrenoyi, 36, 38, 123, 124
  — correlation, 21
  — fauna, 58
  — zone, 35, 36, 37, 38
  "*furcata*," zone, 31
— Gargasian, 23
  joserita, 12, 13, 15, 18, **125,** 153; Pl. 23, figs. 4–6
  — distribution, 13
  — range, 18
  justinae, 22, 36, 39, **124,** 125, 151; Pl. 21, figs. 11–17
  — distribution, 13
  — range, 18
— lower limit in section, 12
  subfurcata, 125
— terminal beds, 50
— time, 51, 57
  truncata, 124
— zone, 50
Dufrenoya?, 152, 153
  compitalis, **125,** 152; Pl. 22, figs. 1–8; Pl. 23, figs. 1–3
  — distribution, 13
  — in Cholla member, 16
Dumble, E. T., Bisbee beds described, 4
— on Bisbee Trigoniae, 5
— on Malone Trigoniae, 6
Durango, 52
— eastern, 50

Eastern communication, 58
Edwards limestone, 5
El Paso, 57
El Tigre, Sonora, 51
— area, 52
Empire Mountains, 60
Enallaster, sp., 28
England, southeastern, 56
Epiaster, sp., 29
Eriphyla, 42
Espinal grit, 11
— large ammonites in, 15
— near shore line, 15
Euhoplites lautus, zone, 36, 38
Eurasian affinities, new, 1
— ammonites, 2
Europe, continental, 56
— eastern, 56
Excentrica group, 40, 86, 88
— present in Arkill, 15
Excentricae, 86
Exogyra, 140
  arietina, 53
  — in Huachuca Mountains, 30
  boussingaultii, 60
  laevigata, 66
  lancha, 22, 56, **66**, 67, 140; Pl. 10, figs. 4–6
  — distribution, 13
  — in Lancha, 14
  minos, 66
— Say, 66
— sp., 66, 140
  tuberculifera, 66

Faecal pellets, 63
Finlay, Texas, 41
Fish Canyon, 60
Flores, T., cited, 52
Folkestone Gault, 14
Forbes, E., 132
Fort Buchanan formation, 59
Fort Crittenden formation, 59
Frech, F., 132
Fredericksburg, 31, 53
— Albian age, 5
— division, 5
— horizons, 5
Fritel, P. H., 32, 132
furcatus?, zone, 33, 34

Galliher, E. W., 63, 132
Gargasian, 36
Gerhardt, K., 63, 132
Germany, northern, 31

Gervillia, 15, 16, 17, 64, 139, 140
  alaeformis, 64
  cholla, 22, 56, **64**, 140; Pl. 10, fig. 1
  — distribution, 13
  — Cretaceous, 64
  — Defrance, 64
  enigma, 64
  heinemani, 22, 56, **64**, 139; Pl. 9, figs. 6–7
  — distribution, 13
  rasori, 16, 22, 56, **65**, 140; Pl. 10, figs. 2–3
  — distribution, 13
Gibbosa group, 86
Gibbosae, 86
Gillet, S., 66, 83
Gilluly, J., cited, 29
Glance conglomerate, 24, 26, 29
— age, 23
— correlated with Trinity, 5
— Morita sequence, as Neocomian-Aptian, 5
— original description, 4
— structure, 7
Glauconia branneri, 5
— aff. *branneri*, 5
"*Goniomya calderoni*," 82
Gorgosaurus libratus, 59
Government Draw, 29
Grammatodon, 16
  sp., 9
Grayson (Del Rio) formation, 31
Grayson-Del Rio, Texas, equivalent of, 1
Greaterville fault, 60
Guadalupe Canyon, 28, 29
— Mountains, 28

Hacienda Saucillo, 51
Hall Ranch, 28
hambrovi, subzone, 34, 35
Hatchet Mountains, 57
Hauterivian, 42, 43
Hayden, F. V., 133
Hay Flat, 28
Heineman, R. E., 3, 63
Hill, R. T., 88, 89, 125, 132
— on basal Comanchean conglomerate, 40
hillsi, subzone, 34, 35
Homomya?, 16
  solida, 21
Hoplites dentatus, 36, 37, 38
"*Hoplites*" *scioptychus*, 127
Huachuca Mountains, 29, 30, 53
— section, 30
— top, 30
Hudspeth County, Texas, 41
Hyatt, A., 48, 104, 132

Hyatt, A. (Cont'd.)
— cited, 96
Hypacanthohoplites, 32, 33, 95, 106, 119, 120
— genotype, 32
  jacobi, zone of, 1, 21
Hypacanthohoplites?
  jacobi, 36, 38
  — zone, 21
  milletianus, 121
"*Hypacanthohoplites*" *jacobi*, 32
Hyphoplites campichei, 115
Hythe beds, 91

Idoceras clarki, 42
  schucherti, 45
Idonearca, 63, 138, 139
  brevis, 63
— Conrad, 63
  stephensoni, 22, 56, **63**, 138, 139; Pl. 8, figs. 9–12; Pl. 9, fig. 1
  — distribution, 13
  — in Lancha, 14
Imlay, R. W., 3, 45, 51, 132
— on *Exogyra arietina*, 30
— on Sonora paleogeography, 52
Immunitoceras, 95, 109, 117, 150
— gen. nov., 116
  immunitum, 9, 22, 36, 54, 109, 116, 150
  — genotype, **117,** Pl. 20, figs. 8–15
  — distribution, 13
  — range, 18
  nolani, 31, 32, 36, 37, 38
  — var. *crassa*, 117
  — var. *pygmaea*, 117
  — var. *subrectangulata*, 117
  nolani, zone, 1, 21
Immunitoceras? uhligi, 109, 116, 117

Jaanusson, V., 130
Jacob, C., 55, 104, 106, 112, 133
— on *Parahoplites aschiltaensis*, 96
jacobi, subzone, 34, 35
— zone, 23, 33
— of Spath, 32
"*Janira morrisi*," 91
Joserita member, 13, 18, 21, 27, 58
— ammonites in, 14
— analyzed, 15
— described, 11
— Pseudo Quadratae introduced, 15

Kachh (Cutch), 48
Karpinsky, A. P., 48
Karsten, H., 84, **85,** 133

Kazansky, P. A., 35, 36, 54, 55, 105, 116, 124, 133
— on *Deshayesites*, 123
— on *Kazanskyella daghestanica*, 100
— on *Parahoplites*, 96
Kazanskyella, 1, 12, 95, 97, 98, 147
  arizonica, 12, 22, 27, 36, 54, 56, **100,** 147: Pl. 17, figs. 2–8
  — distribution, 13
  — range, 18
  — related to Eurasian species, 14
  daghestanica, 38, 56, 100, 101
— gen. nov., 99
— genotype, 100
Kellum, L. B., 50, 51, 52, 133
Khorasan, 55
Kiangsi, 55
Kilian, W., 39, 96, 123
King, R. E., 133
Kislowodsk, 35, 37
Kitchin, F. L., 43, 46, 48, 61, 68, 70, 80, 81, 133
— evaluation of Trigoniae development, 42, 44
— on Malone Trigoniae, 41, 42
Knight, J. B., 49, 133
Kniker, H. T., 92, 133
Koenen, A. v., 133
Kossmat, F., 133
Kossmatia, 42
  aguilerai, 43
  zacatecana, 43

Lagoona District, Coahuila, 44, 50
Lancha limestone, 12, 14, 55
— base of Lowell formation, 14
— contact with Morita, 12
— gastropods in, 21
Large lamellibranchs, occurrence, 15
Las Vigas, beds, Texas, 23
— formation, 50
Lasky, S. G., 57, 133
Lee, C. A., 3
Lee Siding, 28
Leopoldia, 98
  castellanensis, 97
Leymerie, A., 123, 133
Leymeriella regularis, subzone, 35
  schrammeni, subzone, 35
  tardefurcata, 36, 38
  — fauna, 31
  — subzone, 35
  — zone, 32, 35, 36, 38
Leymeriellan-Hoplitan succession, 31

Lima Bruguière, 92, 93, 139
   cholla, 22, 56, **93**, 139; Pl. 9, figs. 3–4
   — distribution, 13
   depressicostata, 93
   dupiniana, 93
   espinal, 56, **92**, 139; Pl. 9, fig. 8
   — distribution, 13
   — in Espinal grit, 15
   mexicana, 93
   muralensis, 20, 28, **93**, 139; Pl. 9, fig. 2
   subaequilateralis,
   vacoensis, 93
Lima? acutilineata texana, 93
Lisson, C. A., 133
Little Hatchet Mountains, 57
Loping area, 52
Lovelace, M. B., 3, 30, 31
Lowell formation, 1, 8, 12, 13, 14, 17, 18, 19, 20, 23, 24, 26, 27, 29, 58
— Aptian age, 6
— Aptian part, 21, 22
— contents, 8
— description, 8
— exotic fauna, 22
— fossiliferous strata, 19
— lateral variations, 23
— localized, 57
— lower part correlated, 21
— Mural limestone, contact, 18
— oscillatory deposition, 14
— selected section, 8, 19
— stratigraphic position, 6
— structure, 7
— successive ammonite faunas, 21
— type locality, 8, 18
— type section, 3
— uninterrupted sequence, 19
— western limit, 53
Lower California, 52
Lower Cretaceous paleogeography, 53
— sequence, 1, 2
— strata, 3
Lower Greensand, 91
Lucina, 15, 16
Lunatia pedernalis, 5
   sp., 20, 25, 28
Luristan, 55
Lycett, J., 85, 133
Lyelliceratidae, 95, 128

Maldanid worms, 63
Malone, 45, 72, 76, 77, 81, 82
— area, Texas, 6
   — visited, 3

Malone (Cont'd.)
— controversy, 41
— facies, 50, 58
   — in Arizona, 44
   — in Mexico, 44
— Hills, 41
— Mountains, 41
— Texas, 5, 6
— Trigoniae, 6
Mangyshlak, 54, 55
— Peninsula, 17
Martinez Ranch, 60
Mayfield Canyon, 37
Mazapil, 50, 51
— area, Zacatecas, 51
Mazatzal Land, 53
Mearns, E. A., 29
Meek, F. B., 2, 133
melchioris, zone, 31, 35, 37
Meretrix, 16
"*Mesalia seriatim-granulata*," group, 21
Mesozoic land bridge, 54
Mexican Canyon fault, 23
— geosyncline, 50, 52, 56, 58
— sea, 50, 52
— trough, 58
Mexico, 8, 42, 58
— eastern, 56
— northwestern, 58
Miller Canyon, 30
milletianus, subzone, 34
Modiola, 16
Molly Gibson formation, 30, 31, 53, 58
— correlated, 41
— localized, 57
— Mine, 31
Morita, 12, 29
— correlated with Trinity, 5
— formation, 4, 6, 8, 13, 23, 24, 26
   — age, 23
   — angular inclusions in, 12
   — Bedoulian age, 23
   — contact with Lowell, 12
   — Lancha, contact, 24
   — standard unit, 2
   — structure, 7
— maroon shales, 26
— material in limy rock, 12
— mudstone, 24
— resembling Cintura, 29
— rocks, 24
— quartzites, 24
Mortoniceras inflatum, 36, 37, 38
Monument Ridge, 19

Moore, H. B., 63, 133
Montezuma Pass, 30
Mount Ararat, 55
Mule Mountains, 19, 20, 27, 28, 29
— Lowell formation in, 8
— Morita-Lowell contact in, 13
— passage zone in, 18
— sections in, 8, 23
— summit, 23
Muller, S. W., 34, 134
Mural, 18, 29, 57
— correlated with middle Albian, 5
— Hill, 23, 24, 25
— Ridge, 26
— section, 8, 26
— limestone, 6, 8, 17, 19, 24, 25, 26, 27, 28, 29, 30, 53, 58
— above Lowell formation, 8
— absence at Tombstone, 29
— as Aptian, 5
— as caprock, 57
— as stable unit, 57
— basal beds, 20, 24
— base, 26
— contacts, 18
— differentiation, 6
— equivalent to Glen Rose, 1, 23
— original description, 4
— regional deepening, 18
— standard unit, 2
— structure, 7
— type, 18, 26
— upper part, 6
— western limit, 53
— Ridge, 23
— sections, 23

Naco, 8
Nagao, T., 134
Nalchik, 35
Nebrodites nodosocostatus, 43
Neithea Drouet, 91, 141
  alpina, 92
  irregularis, 92
  morrisi, 91, 92
  occidentalis, 91
  quinquecostata, group, 91
  vicinalis, 9, 17, 22, **91**, 92, 141; Pl. 11. figs. 6–7
— distribution, 13
Neocomian-Aptian, 5
Neocomites, 123
  cf. *indicus*, 43
Neocomites? praeneocomiensis, 50

Neoharpoceras hugardianum, 97
Nerinea sp., 9, 36
Neumayr, M., 134
Neuquen area, Argentina, 15
New Mexico, 57
Nichols, T., 3
Nikchitch, J., 104, 105, 134
Nikitin, S., 134
Ninety One Hills, 8, 17, 18, 23, 24
— area, 17, 24, 28
— geological map, 18
— panoramic view, 18
— quartzites, 4, 24
— topography, 3
— in focal point, 7, 8
— section, 8, 19
nodosocostatum, zone, 34
nolani, clay, 31, 32
— subzone, 34
— zone, 37
North America, new faunas for, 1, 2
North Caucasus, 35
Nucula nucleus, 63
— (*Nucula*) *capaensis*, 62
— *nucleus*, 62
  pectinata, 61, 62
Nuculidae, 61, 62
nutfieldensis, subzone, 34

Oomia, 70
Oppelia (*Aconeceras?*) *trautscholdi*, 37, 38
Orbigny, A.d', 93, 104, 120, 121, 134
Orbitolina, 58
— Albian, Glen Rose, 23
— beds in Guadalupe Canyon, 29
  texana, 5, 8, 17, 18, 20, 23, 26, 28, 29, 36, **37**, 40, 57, 58
— beds, 23
— in Mural, 1
— in thin beds, 6
— stratigraphic position, 20
— zone, 17, 22
— zone, 79
Oregon, 54
Ostrea Linnaeus, 16, 66, 141
  cortex, 66
  edwilsoni, 16, 22, 56, **66**, 91, 141; Pl. 11, figs. 1–3
— distribution, 13
  praelonga, 66
  ragsdalei, 40
— occurrence in Arizona, 30
— sp., 5
Ostreidae, 66

INDEX

Pacheta member, 12, 13, 18, 19
— analyzed, 14
— described, 12
— Saavedra members, 1
Pacific-Caucasian faunas, in Arizona, 55
Pacific Ocean, 54
Paleogeography, regional, 50
Paleozoic strata, in Patagonia Mountains, 30, 31
Panope, 17
Panther Creek, Texas, 39
Paracanthohopites, 33, 58, 95, 109, 119, 151
— gen. nov., 118
  meridionalis, 21, 22, 36, 54, 56, 109, **118**, 151; Pl. 21, figs. 1–7
  — distribution, 13
  — genotype, 118
  — of Caucasian relation, 15
  — range, 18
  multispinatus, 56, 109, 119
Parahoplitan fauna in Pacheta, 21
Parahoplites, 37, 52, 56, 95, 98, 107, 108, 121, 124, 147
— Anthula, emend., 99
  aff. *melchioris*, 51
  campichei, 109, 115, 116
  inconstans, 99
  jacobi, 119
  maximus, 98, 103
  melchioris, 36, 37, 38, 56, 98, 99, 100, 101, 116
  — genotype, 99
  — original description, 96
  — zone, 23, 36, 38
  "*melchioris*," 97, 147
  multicostatus, 97
  nutfieldensis, subzone, 35
  stantoni, 54
  subcampichei, 116
  treffryanus, 97
"*Parahoplites*" *aschiltaensis*, 32, 51, 103, 107
  hanovrensis, 33
  jacobi, 32, 33, 120
  cf. *multispinatus*, 51
  treffryanus, 107, 121
  uhligi, 32, 54
"*Parahoplites*" *umbilicostatus*, 122
Parahoplitan-acanthohoplitan succession, 31
Parahoplitan fauna, infiltration, 56
"Parahopliten Schichten," 31
— of Stolley, 32
Parahoplitidae, 32, 49, 95
— analysis, 49
— critical revision, 2

Parahoplitinae, 31, 49, 95, 108, 123
— costation, 97
— suture, 98
Parahoplitoides, 123
Parallelodontidae, 63
Parker Canyon, 30
Patagonia, Arizona, 59
— Cretaceous of, 15, 30
— group, 30, 31, 40, 53, 58
— Mountains, 6, 29, 40, 53, 58, 59
  — ammonites from, 5
— South America, 80
Paul Spur, 6, 26, 28
Pecten Osbeck, 15, 16, 17, 21, 90, 146, 147
  (*Chlamys*) *elongatus*, 90
    sp., 27
    thompsoni, 16, 22, 56, **90**, 146, 147; Pl. 16, figs. 5–8; Pl. 17, fig. 1
  elongatus, 90
  quadricostatus, 91
  stantoni, 5
  thompsoni, 11, 91
Pectenidae, 90
Pedregosa member, 17, 18, 24, 27, 28, 29
— analyzed, 17
— contents, 8
— described, 8
— in Guadalupe Canyon, 29
— Mountains, 28
— Mural, contact, 18
  — passage zone, 18
— strata, gastropods in, 9
— type, 26
Pentagonaster? sp., 31
Perilla member, 13, 17, 22, 24, 28, 29, 58
— analyzed, 17
— contents, 9
— described, 9
— in Guadalupe Canyon, 29
— Mountains, 28
— Texas fauna in, 1
— yellow shale in, 29
Permian limestone, 30
— time, 55
Perna, 15
  militaris, 65
Pernidae, 64
Peru, 44
Peterson, N. P., 3
Pholadomya lerchi, 40
Physa, 59
Pictet, F. J., 93, 134
Pleuromya, 44
Pleuromya?, 16

Plicatula, 16
Pompeckj, J. F., 2
Portugal, 66
Potrillo Mountains, 57
Protocardia, 17, 18
Pseudo-Quadratae, 67
— age in western hemisphere, 48
— important modifications, 46, 69
— in Arizona Aptian, 1, 75, 76, 77, 78
— in Sonora, 52
— Kimmeridgian in Texas, 1, 41, 43, 44, 46, 48, 70, 71, 72, 73, 74
— Neocomian in Argentina, 1, 46
Pseudo-quadrate Trigoniae, 10, 11, 16, 44, 56
Pteria Scopoli, 65, 139
  carteroni, 66
  cottaldina, 65
  peregrina, 17, 22, 56, **65**, 66, 139; Pl. 9, fig. 5
  — distribution, 13
Pteriidae, 65
Ptychomya, 41, 42
  stantoni, 42

Quadratae, 47, 67, 68, 69, 70, 73, 76, 78, 79
Quajote member, 1, 9, 13, 18, 58
— acanthohoplitan fauna in, 1, 21
— ammonites of, 14
— analyzed, 16
— described, 9
— exotic fauna in, 16
Quenstedt, W., 61, 62, 134
Quimbo dolomite, 11, 15
Quintuco formation, 65
Quitman Mountains, Texas, 17, 23, 57, 58
— area, 3, 45
— Cretaceous sequence, 37
— sections, 3
— stratigraphy, 22

Radiolites? sp., 20, 28
Ransome, F. L., 134
— Bisbee group introduced, 4
— cited, 29
— collections from Bisbee, 90
— geological map, 7
— on ammonites, 6
— on Bisbee anticline, 7
— on collecting localities, 6
— on granite porphyry, 26
— on Mural limestone, 6
— on porphyritic rocks, 26
— on *Trigonia stolleyi*, 6
Rasor, C. A., 29, 131

Reeside, J. B., Jr., 3, 81, 86
— cited, 79
regularis, subzone, 34, 35
Renevier, E., 134
Renngarten, V. P., 35, 36, 37, 134
— cited, 116
Reptilian remains, 12
Rhynchonella sp., 5
Riedel, L., 121, 134
Rio Grande, 3, 23, 32
Rio Nazas area, 32
— Durango, 51
— fauna, 32, 51
Roemer, F., 22, 92, 134
Roman, F., 99, 123, 134
— on Cheloniceratidae, 32
Rostellaria?, 15
Rostral dent in *Acila*, 62
rude, subzone, 34

Saavedra member, 1, 13, 16, 21, 24, 27, 58
— analyzed, 15
— described, 11
Salenia sp., 16
Saltillo, Coahuila, 51
San Bernardino, 57
San Carlos Mountains, Tamaulipas, 51, 52
Sandberg, C. H., 3
Sandy Bob Ranch, 29
Santa Rita Mountains, 59
Sarasin, C., 98, 134
Sarstedt, 31
Scabrae, 88
Scabroid Trigoniae, 11
Schenck, H. G., 14, 34, 55, 56, 61, 62, 134
Schlaudt, A., 3
— Ranch, 18, 19
— Ridge, 19
Schrader, F., 5, 31, 59, 134
schrammeni, subzone, 34
Schuchert, C., 134
Scott, G., 2, 37, 45, 57, 128, 135
— *Cheloniceras* described, 32
— collections, 3
— on *Leymeriella regularis* zone, 39
Semilaeves, 86
"*Serpula*," 10, 16
Seunes, J., 116, 135
Sharpe, D., 135
Shasta County, California, 54
— series, 54
Sierra Blanca, Texas, 5, 41

Sinzow, J., 17, 31, 32, 35, 54, 55, 56, 98, **116,** 135
— on *Acanthohoplites*, 106
— on *Parahoplites*, 96
Sinzowiella, 1, 97, 98, **116,** 148
— gen. nov., 101
— genotype, 101
spathi, **102,** 103, 148; Pl. 18, figs. 1–17
— distribution, 13
— range, 18
— related to Eurasian species, 14
Sinzowiella? 153
— sp., **103**; Pl. 23, figs. 7–13
— distribution, 13
Smith, J. P., 50, 135
Sonneratia, fauna, 37, 39
rossica, 103
trinitensis, 36, 38
— zone, 22, 23, 35, 39
Sonoita Flat, 59, 60
— group, 59
Sonora, central, 52
— northeastern, 58, 60
Sonoran trough, 53
South Africa, 42, 70
South America, 42, 43, 70
— northern, 56
South American faunas, 55
South Andean province, 69
— realm, 42
sparsicosta, subzone, 34
Spatangidae, 10
Spath, L. F., 32, 34, 43, 48, 67, 97, 104, **116,** 121, 123, 124, 128, 135
— cited, 50, 67
— on Cheloniceratidae, 32
— on Malone ammonites, 41
Sphaerium, 60
Stanton, T. W., 5, 54, 59, 104, 135
— on *Trigonia stolleyi*, 6
— on "*Pecten stantoni* (Hill)," 90
Stanton's topotype from Truncate Mound, 47
Stein, B., 3
Steinmann, G., 67, **135**
— on Trigoniae, 46
Stenhoplites, 123
Stenzel, H. B., 3, 21
Stephenson, L. W., 3, 93, 135
Stoliczkaia, 1, 53, 58, 95, 128
dispar, 36, 38, 128
— zone, 31, 36, 38
dorsetensis, 128, 129

Stoliczkaia (Cont'd.)
excentrumbilicata, 36, **129,** 156; Pl. 26, figs. 5–6
— in Patagonia group, 30
— Mountains, 6
— Neumayr, 128
— new forms, 31
patagonica, 36, **128,** 129, 156; Pl. 26, figs. 3–4
scotti, 36, **129**; Pl. 26, figs. 7–8
— in Patagonia Mountains, 2
— zone, 53
Stolley, E., 31, 32, 55, 135
— on Clansayes, 31
Stoyanow, A., 6, 135, 136
— on Patagonia group, 30
Stoyanow, V., 3
subnodosocostatum, zone, 34
Sycamore Creek, Texas, 28

Takahashi, 63
Taliaferro, N. L., 135
"Tardefurcata Schichten" of Germany, 32
tardefurcata, zone, 34
Tenney, J. B., 135
Tethyan-Atlantic, connection, 58
— faunas in Arizona, 55
Tetragramma? sp., 16
Texas, 22, 32, 42, 56, 58
— ammonite fauna of, 1
— Cretaceous strata, 21
— East Mexican faunas in Arizona, 55
— region, 50
— southwestern, 22
— species, gradual infiltration, 21, 56
— stratigraphic units, correlated, 40, 41
Thetironia caucasica, 38
Thompson, F. O., 2
Tobler, A., 133
Tombstone, 29, 53
— Cretaceous outlier, 29
— district, 29
— Mesozoic section, 29
tovilense, subzone, 34
Transcaspian Region, 2, 17, 35
trautscholdi, subzone, 34
Travis Peak, 37, 39
— formation, 39
Trigonia, 17, 67, 142, 143, 144, 145, 146
— Bruguière, 67
abrupta, 83, 84, 86
— group, 83
agrioensis, 86, 87
aliformis, 83, 88, 89

Trigonia Cont'd.)
— group, 88
calderoni, 42, 44, 47, 81, 82, 144
— collected by Stanton, 81; Pl. 14, figs. 1–2
— topotypes, 45
caudata, 88, 89
conocardiiformis, 80, 81, 82
cragini, 10, 16, 22, 47, 48, **80,** 81, 82, 143, 144;
   Pl. 13, figs. 6–10; Pl. 14, fig. 3
— distribution, 13
— range, 18
— upper limit, 12
crenulata, 88
daedalea, 79
dubia, 79
dumblei, 46, 70, **77;** Pl. 12, fig. 4
dunscombensis, 87
emoryi, 90
excentrica, 87
— group, 86
guildi, 16, 22, 46, 47, 69, **70,** 73, **75,** 142; Pl. 12, figs. 1–2
— distribution, 13
— range, 18
herzogi, 68, 69, 75
heterosculpta, 80
holubi, 68
hondaana, 76, 83
humboldi, 83, 84, 85
kitchini, 10, 16, 22, 48, 80, **82,** 83, 144; Pl. 14, figs. 4–10
— distribution, 13
— range, 18
— upper limit, 12, 16
laeviuscula, 87
lerchi, 86
lingonensis, 83
lorentii, 83, 85
maloneana, 74
— var., 74
mammillata, 68, 70
mearnsi, 1, 9, 17, 22, 27, 29, 30, 37, 39, 47, 57, 70, **78,** 79, 89, 99, 142, 143; Pl. 12, figs. 5–6; Pl. 13, figs. 1–4
— distribution, 13
— range, 18
— type locality, 29
— zone, 17, 23, 39, 57
meyeri, 85
nodosa, 79
picunensis, 81
proscabra, 42
recurva, 80

Trigonia (Cont'd.)
reesidei, 12, 19, 22, 27, 40, 58, **83,** 84, 85, 86, 144, 145; Pl. 14, figs. 11–14; Pl. 15, figs. 1–3
— distribution, 13
— in Espinal grit, 15
— in Lancha, 14
— last recurrence, 15
— range, 18
resoluta, 16, 46, 70, **76,** 77, 142; Pl. 12, fig. 3
— distribution, 13
— range, 18
rogersi, 83
saavedra, 23, 47, 70, **78,** 143; Pl. 13, fig. 5
— distribution, 13
— range, 18
— sp. ex *aliformis* group, **89,** 146; Pl. 16, figs. 1–3
— distribution, 13
steeruvitzii, Dumble on, 5
"*steeruwitzi*," Adkins on, 5
stolleyi, 1, 5, 6, 17, 22, 23, 39, 57, **88,** 89, 90, 145; Pl. 15, figs. 9–11
— distribution, 13
— in Glen Rose, 22
— in Guadalupe Canyon, 29
— range, 18
sulcataria, 83
taffi, 77, 79
— auctorum, 1
— Cragin, misinterpreted, 46
tocaimaana, 90
transitoria, 46, 68, 69, 73, 75
— group, 43
— paired costallae in, 69
— var. *curacoensis*, 46
— var. *quintucoensis*, 46, 69
— var. *vacaensis*, 46
undulata, 83
v-scripta, 79
vau, 79, 80
vyschetzkii, 41, 42, 43, 44, 46, 47, 69, 70, 71, 75, 77
— Cragin, emend., 73
— Cragin, lectotype, 45
— Cragin (part), 72
— status, 70
— topotypes, 45
weaveri, 16, 19, 23, 86, **87,** 88, 145; Pl. 15, figs. 4–8
— distribution, 13
— range, 18
Trigoniae, 6, 16

Trigoniae (Cont'd.)
— of *abrupta* and *excentrica* groups in Arizona, 21
— of *aliformis* group, 21
— in Sonora, 52
— Cretaceous of Jurassic aspect, 2
— Cretaceous, in Texas, 21
— Jurassic, at Malone, 2, 21
— pseudo-quadrate, 21, 27, 28
— in Sonora, 44
— scabroid, 27, 28
— of *v-scripta—vau* groups, 1, 21, 44, 48
trinitensis, zone, 23
Trinitoceras, 22, 57, 79
— below *Orbitolina texana* zone, 17
rex, 37, 57
Trinity division, 5
— group, Texas, 5
Tropaeum bowerbanki, subzone, 35
Tschernyschewia, 55
Tucson, 1, 5, 29
Turitella sp. cf. *seriatim-granulati*, 5
Tusonimo linestone, 12
— recurrence of ammonites, 14

Uhlig, V., 43, 134, 136
— on Malone age, 41
— on similarity of Trigoniae, 44
"*Uhligella*" aff. *walleranti*, 51
Umia, 70
Unicardiidae, 95

Unicardium, 15, 16, 95
— sp., 9, **95,** 146; Pl. 16, fig. 4
Unio, 59, 60
Upper Cretaceous deposits, 59

V-Scripta—vau groups, 42, 79
— interpretation of species, 47
Valanginian, 42, 43
Venezuela, 56
Viviparus, 59
Vola irregularis, 91

Washita, 53
— time, 31
Weaver, C. E., 43, 65, 81, 87, 136
— on Argentina Trigoniae, 42, 46
weissi, zone, 34
Weller, S., 91, 136
Western communication, 58
Whetstone Mountains, 60
Whitfield, R. P., 91, 136
Whitney, F. W., 3
Wilson, E. D., 3, 6, 29, 131
Winton, W. M., 131
Winwood addition, 23, 24
Woodring, W. P., 28, 136
Woods, H., 61, 62, 136
— on *Acila*, 14

Yagi, 63

Zittel, C., 96